*L. Guglielmo · A. Ianora (Eds.)*
*Atlas of Marine Zooplankton / Straits of Magellan*

*Amphipods, Euphausiids, Mysids, Ostracods, and Chaetognaths*

# Springer

*Berlin*
*Heidelberg*
*New York*
*Barcelona*
*Budapest*
*Hong Kong*
*London*
*Milan*
*Paris*
*Santa Clara*
*Singapore*
*Tokyo*

L. Guglielmo    A. Ianora    (Eds.)

# Atlas of Marine Zooplankton Straits of Magellan

## Amphipods, Euphausiids, Mysids, Ostracods, and Chaetognaths

T. Antezana, G. Benassi, G. Costanzo, N. Crescenti, I. Ferrari,
E. Ghirardelli,  A. Granata, L. Guglielmo, K.G. McKenzie

With 124 Figures

 Springer

Professor Dr. LETTERIO GUGLIELMO
Università degli Studi di Messina
Dipartimento di Biologia ed Ecologia Marina
Salita Sperone 31
98166 Messina-S. Agata
Italia

Dr. ADRIANNA IANORA
Stazione Zoologica Anton Dohrn
Villa Comunale
80121 Napoli I
Italia

L. Guglielmo/ A. Ianora (Eds.):

**Atlas of Marine Zooplankton / Straits of Magellan / Copepods (1995)**
ISBN 3-540-58228-2

**Atlas of Marine Zooplankton / Straits of Magellan**
**Amphipods, Euphausiids, Mysids, Ostracods, and Chaetognaths (1997)**
ISBN 3-540-58678-4

ISBN 3-540-58678-4 Springer-Verlag Berlin Heidelberg New York

Library of Congress Cataloging-in-Publication Data. Atlas of marine zooplankton, Straits of Magellan / L. Guglielmo, A. Ianora (eds.). p. cm. Includes bibliographical references (v. 1, p.). Contents: [1] Codepods. ISBN 3-540-58228-2 (Springer-Verlag Berlin Heidelberg New York). - 1. Marine zooplankton-Magellan, Straits of (Chile and Argentina) – sAtlases. I. Guglielmo, L. Letterio), 1945-. II. Ianora, A. (Adrianna), 1953–QL 123.A85 1995  592.09294–dc20  94-39998

Cover design: E. Kirchner, Heidelberg
Typesetting: perform, k + s textdesign GmbH, Heidelberg
SPIN 10488056  31/3137  5 4 3 2 1 0 - Printed on acid-free papier

In memory of the late Professor Sebastiano Genovese,
Director of the Stazione Zoologica and
Head of the Marine Ecology Department
of the University of Messina

Italian Ministry of the University and Scientific and Technological Research
National Research Program for Antarctica (PNRA)

# Foreword

One is struck by the fact that, ever since Magellan himself passed through the straits now bearing his name, countless ships of dicovery, exploration and scientific investigation have viewed this South American thoroughfare as a point on the way rather than a destination. Beginning with Captain James Cook, who in the 1760s firmly established the practice of carrying scientists abroad, continuing with Darwin in the 1830s, and even to the present day, ships bearing scientists through the Straits of Magellan have generally been in transit to some other more tantalizing locale. It is surprising that this body of water, representing the transition between temperate and Antarctic, Atlantic and Pacific Oceans, has received so little attention. Thus, the modern research of the plankton fauna of this region by Italian scientists is all the more remarkable for being the first to adequately recognize and to study the organisms in such detail.

The first fruit of these labors was an authoritative tome on the Copepoda which, in addition to its clarity of presentation, described 18 species new to the region. The present volume, covering the chaetognaths, amphipods, ostracods, euphausiids and mysiids, is no less of an accomplishment. Among these are several species already well known in the Antarctic, such as *Sagitta maxima, Eukrohnia hamata, Orchomene plebs, Primno macropa and Themisto gaudichaudi.* Those who have been accustomed to traveling further south to study these species might now consider the advantages of being able to carry out some of their research without having to run the gauntlet of the infamous Drake Passage. On the other hand, comparative studies of these species in both Antarctic and sub-Antarctic provinces may yield substantial insight into the processes underlying their biogeography.

Before we can begin to ask sensible questions about how a pelagic community functions, we must first know of which species it is composed. This volume, along with its copepod companion, now sets the stage for the studies of ecosystem, community and population dynamics that will challenge future generations of marine scientists.

Scripps Institution of Oceanography                                    Mark E. Huntley
La Jolla, California, USA

During the past two decades, research on the biology of the Antarctic has focused mainly on the Antarctic region itself. Thus, regional faunal studies in sub-Antarctic waters, as carried out during the course of the Italian Expedition in the Straits of Magellan in 1991, are of great scientific interest.

The Straits, belonging to the Magellan Province, are well known for their relatively high diversity of species; but only very few detailed faunistic surveys have been carried out so far. Both the oceanographic properties and the biota are essentially influenced by the topography of the Straits. The faunal assemblages include species that are typical for Antarctic and sub-Antarctic areas, as well as those that are more cosmopolitan. In other respects, the fauna of the Straits shows major links with the pelagic fauna of temperate regions.

The *Atlas of Marine Zooplankton, Straits of Magellan* is an important contribution to the knowledge of the plankton communities in sub-Antarctic regions. While the first volume dealt with the taxonomic composition and ecology of the copepods, the world's most numerous group in the pelagic community, the present volume describes those groups of pelagic organisms that rank next in abundance: amphipods, ostracods, chaetognaths, euphausiids and mysids. Several of these species reported have not been previously sampled in this geographic area.

The primary concern of the Atlas is to facilitate and promote future biological studies in the Antarctic by both experts and beginners. Additionally, the handbook provides a broad basis of knowledge that will equip the reader with a greater understanding of plankton communities and the biogeography of this sub-Antarctic region.

Universität Hamburg                                        Hans Georg Andres
Zoologisches Institut und Museum
Hamburg, Germany

# Preface

The second Atlas of the zooplankton assemblages of the Straits of Magellan is intended as a companion text to the first volume on the copepod species in this biogeographically unique area. The Atlas deals with the amphipods, euphausiids, mysids, ostracods and chaetognaths sampled during the course of the Italian expedition in the Straits in 1991. Although numerically less abundant than the copepods described in the first volume of this series, these other groups can at times be relatively abundant, ranking second or third in frequency after the ubiquitous copepods. Also, they contribute significantly to the total zooplankton biomass and therefore play an important role in the carbon flux dynamics in the area.

The Atlas is intended for both beginners and experts in plankton ecology. The descriptive format is similar to the one adopted for the first volume on copepods. It is extremely succinct, with brief taxonomic descriptions. Emphasis is given to illustrations that include line drawings and scanning electron micrographs; 39 species are described including 13 amphipod, 6 euphausiid, 4 mysid, 13 ostracod and 5 chaetognath species. Several of these are new for the Straits. As in the first volume of the series, detailed information is also given on species distributions and relative abundances, together with environmental data at the time of collection. A detailed description of the study area has already been published in the first volume of the series.

Since much of the faunistic information on the zooplankton assemblages in sub-Antarctic areas dates back to the turn of the century and is usually contained in rare monographs or reports, we hope that this new Atlas, together with the first volume, will provide a rapid and practical handbook for the easy identification of the species living in this and other sub-Antarctic areas, facilitating future research activities in these regions.

# Acknowledgements

Once again, it is a pleasure to record our gratitude to those who have contributed to the preparation of this new Atlas. As for the first volume, Giorgio Dafnis of the Stazione Zoologica prepared the material for the scanning electron microscope and took all of the SEM micrographs, with the assistance of Gennaro Iamunno. Giuseppe Gargiulo printed the micrographs. Pino Arena generated the distribution maps and Vincenzo Bonanzinga helped in the data analysis. In addition, we thank the crew of the R/V *Cariboo* who participated in the 1991 cruise to the Magellan Straits, and the National Antarctic Research Program (PNRA) for their financial support.

We are grateful to Dr. H.G. Andres (Hamburg University) and Prof. S. Ruffo (Museo Civico di Storia Naturale, Verona) for their critical comments and suggestions on the chapter regarding amphipods, and to Dr. A. Alvariño (National Marine Fisheries Service, La Jolla) for use of the drawings on chaetognaths.

# Contents

T. Antezana
University of Conception
Oceanography Department
Pelagic Ecology Laboratory
Casilla 2407
Concepcion, Chile

G. Benassi
Department of Environmental Sciences
University of Parma
Viale Delle Scienze
43100 Parma, Italy

G. Costanzo
University of Messina
Faculty of Science
Department of Animal Biology
and Marine Ecology
Contrada Sperone 31
98166 S. Agata Messina, Italy

N. Crescenti
University of Messina
Faculty of Science
Department of Animal Biology
and Marine Ecology
Contrada Sperone 31
98166 S. Agata Messina, Italy

I. Ferrari
Department of Environmental Sciences
University of Parma
Viale Delle Scienze
43100 Parma, Italy

E. Ghirardelli
Biology Department
University of Trieste
Via E. Weiss 2
34127 Trieste, Italy

A. Granata
University of Messina
Faculty of Science
Department of Animal Biology
and Marine Ecology
Contrada Sperone 31
98166 S. Agata Messina, Italy

L. Guglielmo
University of Messina
Faculty of Science
Department of Animal Biology
and Marine Ecology
Contrada Sperone 31
98166 S. Agata Messina, Italy

K. G. McKenzie
University of Melbourne
Parkville
Victoria 3052, Australia

The study area lies within the geographical range of the so-called antiboreal South American (Eckman 1953), Magellanic (Balech 1954; Schilder 1956; Briggs 1974) or Magellanean (Antezana 1981) Province which extends from 42°S to the Antarctic Convergence along the channels of western South America, around its tip to Peninsula Valdes and along the east coast of South America, including the Falkland (Malvinas) Islands. The northern limits and districts of this Province, on the Pacific side, are not well established, according to authors who have studied the littoral (Hedgpeth 1969) and pelagic (Antezana 1981) faunas in this area. Based on the littoral fauna, Balech (1954) distinguished two districts: the *chiloensis* from 41°S to 51°S and the *fuegensis* from 51°S to the tip of South America. Antezana (op. cit.) recognized these two districts for the pelagic fauna as well, but proposed boundaries from 42–47°S and 47–60°S. The break between the two districts was given by the southern limit of the sub-Antarctic temperate euphausiid *Euphausia lucens* and by the northern limit of intrusion of transition zone species such as the euphausiids *Thysanoessa gregaria* and *Nematoscelis megalops* in coastal waters. The southern limit, at the Antarctic Convergence, was associated with the distribution of many planktonic species such as the medusae *Cosmostirella davisi* and *Phialidium simplex,* the salp *Salpa thompsoni,* the pteropods *Limacina helicina* and *Clio sulcata,* and the euphausiids *Euphausia vallentini, E. similis* and *E. longirostris.* The northern limit, which extends up to ca. 38°S, was associated to the limits of most sub-Antarctic species in the Humboldt Current System, such as *Sagitta gazellae, C. davisi, E. vallentini* and *E. lucens.* No districts within the Magellanic Province were recognized by Eckman (1953), Knox (1960), Briggs (1974), and Brattström and Johanssen (1983).

The biota of the fjords, the Straits of Magellan included, includes, as expected, components of the Magellanic Province, such as sub-Antarctic copepods (Marin and Antezana 1985; Mazzocchi et al. 1995), euphausiids (Antezana 1976) and other zooplankters (Guglielmo et al. 1991; Antezana et al. 1992), but also transition zone species (sensu Brinton 1962; Johnson and Brinton 1963) such as the euphausiids *T. gregaria* and *N. megalops* (Antezana 1976).

The intrusion or colonization of the pelagic environment of the Straits by oceanic sub-Antarctic and transition zone species may be limited by physical barriers such as the shallow shelf at the Atlantic entrance and the shallow sill at the Pacific entrance. This may be particularly true for meso- and bathypelagic species, which are often strong diel vertical migrators. Other specific adaptations to an extremely heterogeneous, and physically and chemically modified sub-Antarctic environment may determine the species richness, diversity, dominance and species role in the energy flow within the community. A trend towards strong geographical patchiness of zooplankton with the aggregation of single species (Antezana 1976), high frequency of occurrence of red tides (Clement and Guzman 1989), dominance of *E. vallentini* among the euphausiids (Antezana 1981) have been reported in the context of pelagic adaptation to this estuarine environment. Species differentiation with incomplete isolation may be presently occurring in this geologically new environment.

A major source of discussion refers to the geographical extent of the influence of Atlantic and Pacific waters and fauna in the Straits of Magellan. Based on temperature-salinity diagrams, Pickard (1973) suggested that the change in water mass structure from the Atlantic to the Pacific occurs between Carlos III Island and

Cape Froward in the Intermediate Passage. Artegiani et al. (1991) showed the distribution for temperature, salinity and density in a transect through the Straits. It seems that the major inflow of water and plankton into the Straits comes from the Pacific entrance and from the southern entrance of Magdalena Channel. The influence of Atlantic inflow is restricted to the Second Narrow of the Straits due to the extreme depth and restriction at this point and at the First Narrow. Also, there is a strong discontinuity in temperature as one approaches the eastern entrance (Artegiani and Paschini 1991) and another discontinuity in temperature and salinity at the southern entrance and through the Magdalena Channel (Artegiani et al. 1991). Finally, there is a strong break in the plankton community in the First and Second Narrows (Guglielmo et al. 1991). Consistent with these findings is the presence of the euphausiid *Thysanoessa gregaria* and the absence of another species of euphausiid, *T. vicina*, frequently encountered in the Atlantic sector of the Straits (Antezana and Brinton 1981), suggesting that the inflow of Atlantic waters is mainly limited to the First Narrow.

As already reported in the first volume of this Atlas, there are very few studies on the zooplankton of the Straits. The first volume described the copepod assemblages of the region during the austral summer (February-March), 1991. Of the other zooplankton groups, amphipods, euphausiids, mysiids, ostracods and chaetognaths were the most plentiful in this period and are described in this volume.

With regards to the amphipods, six species of gammariids and seven species of hyperiids are described in Chapter 3. Other species of gammariids and caprelliids may have been present at the time, at stations north of Punta Arenas near the entrance to the Atlantic Ocean. However, these are not mentioned in this volume, because the specimens were either too young to be correctly identified or greatly damaged. Of the 13 species described, only the hyperiid *Themisto gaudichaudii* was sampled in high numbers (4551 specimens, 1447 of which juveniles) at all stations in the Straits. Four species were very rare (*Thryphosella schellenbergi, Jassa falcata, Hyperoche mediterranea* and *Hyperiella dilata-*

*ta*) with only one specimen sampled for each species. In general, hyperiids were more commonly sampled at Pacific stations with the exception of *Hyperoche medusarium,* which was sampled only in Central waters from Cape Froward to Paso Ancho, and *T. gaudichaudii,* present throughout the Straits. By contrast, gammariids were more uniformly distributed from the Pacific entrance up to the Primera Angostura (St. 22).

Six species of euphausiids are described in Chapter 4, *Euphausia vallentini, E. lucens, Thysanoessa gregaria, Nematoscelis megalops, Stylocheiron longicorne* and *S. maximum.* Of these, the most abundant by far was *E. vallentini.* The number and size of individuals of this species increased from the Pacific towards the Atlantic, up to the Paso Ancho area where the largest specimens were sampled. In this area, *E. vallentini* formed dense swarms in proximity to the bottom during the day, and underwent strong vertical migrations at night. *N. megalops* was also sampled throughout the Straits but in lower numbers compared to *E. vallentini.* It was more commonly found in the Pacific area, and also in the Paso Ancho region, where the greatest number of adults was recorded. In contrast, *E. lucens* was sampled mainly at the Pacific entrance, until Tamar Island, and was only sporadically present in the Atlantic sector (St. 26). The three other species were very rare.

Of the four species of mysids described in Chapter 5, *Boreomysis rostrata* was by far the most abundant, accounting for more than 93 % of the species sampled. The species was mainly sampled in the Pacific and Central areas of the Straits, where both sexes and all developmental stages were present. In the Pacific area, it was sampled from Tamar Island to Carlos III Island (Sts. 6, 7, 9, 10, 11), whereas in the Central region it was found, in great numbers, from Cape Froward to Cape S. Isidro (Sts. 13, 14 and 15). Of the three other species of mysids, only two specimens of *Pseudomma magellanensis* were sampled in the Paso Ancho area, and only five adult females of *Arthromysis magellanica* were found at St. 26 in Atlantic waters beyond the Straits. *Neomysis monticelli* was also sampled only at St. 26, where 4 males, 2 females and ca. 100 juveniles were collected.

For ostracods described in Chapter 6, highest densities were recorded in the regions of Cape Froward, Baia Inutil and Paso Ancho which are areas of high phytoplankton biomass and production (Guglielmo et al. 1993). In many samples, densities were greater than 100 ind m$^{-3}$, with a maximum of > 600 ind m$^{-3}$, mainly due to the presence of one species, *Discoconchoecia elegans*. Much lower densities were recorded in the Atlantic sector of the Straits, where one species, *Pseudoconchoecia serrulata*, dominated over the others, and in the Pacific area characterized by a high species diversity.

Of the eight species of Halocyprida sampled in the Straits, three (*D. elegans, P. serrulata and Obtusoecia antarctica*) had already been recorded during the course of the Italian expedition in the Southern Ocean in 1989–1990 (Benassi et al. 1992). In the case of two other species of Halocyprida (*Loricoecia loricata* and *Paramollicia rhynchena*) and the myodocopide *Macrocypridina poulseni*, the present records are the first for the region, representing the most southerly yet for these species.

With regards to the biogeography of the Halocyprida sampled, four species (*Mikroconchoecia acuticosta, D. elegans, L. loricata* and *P. rhynchena*) are transoceanic, reaching as far as 60°N; the four other species (*Metaconchoecia australis, Conchoecilla chuni, P. serrulata* and *O. antarctica*) are confined to the Southern Hemisphere. The northernmost record of the myodocopide *M. poulseni* is 37°N. Four other species are considered epibenthic (*Philomedes eugeniae* and *P. cubitum*) or planto-benthic (*Paradoxostoma* sp. aff. *hypselum* and *P. magellanicum*). Of these, *P. hypselum* is circum-Antarctic in distribution, whereas the others are typical of the Magellanic region of the sub-Antarctic.

Chapter 7 describes the chaetognaths of which *Sagitta tasmanica* was by far the most common, representing 75 % of all specimens collected. This percentage was even higher (95 %) at the Pacific entrance to the Straits. It was the only species sampled throughout the study area. The four other species were much less common: *S. decipiens* (4.5 %), *S. gazellae* (0.8 %), *S. maxima* (1 %) and *Eukrohnia hamata* (12 %). The number of specimens diminished from the Pacific to Atlantic entrances and only on the Pacific side of the Straits were all species sampled. This was probably due to the deeper depths in this area allowing for strong diel vertical migrations. Only one specimen each of *S. decipiens* and *S. gazellae* was sampled east of Cape Froward, at the entrance to Baia Inutil and within the Bay, respectively. Several specimens were in very poor condition, but their general shape suggests the probable presence of other very rare species such as *Eukrohnia bathypelagica, Sagitta marri* and *S. planctonis*.

## References

Antezana T (1976) Diversidad y equilibrio ecológico en communidades pelágicas. In: Orrego F (ed) Preservación del Medio Ambiente Marino. Inst Est Internac U de Chile, Santiago, pp 40–54

Antezana T (1981) Zoogeography of euphausiids of the South Eastern Pacific Ocean. Memorias del Seminario sobre Indicadores Biológicos del Plancton. UNESCO, Montevideo, pp 5–23

Antezana T, Brinton E (1981) Euphausiacea. In: Boltowskoy D (ed) Atlas del zooplancton del Atlántico Sudoccidental. INIDEP, Buenos Aires, pp 681–698

Antezana T, Dellarossa V, Zúñiga A, Rosas A (1992) Features of the pelagic environment of Chilean fjords. In: Gallardo VA, Ferretti O, Moyano HI (eds) Oceanografia in Antartide. ENEA-PNR, Rome, pp 460–466

Artegiani A, Paschini E (1991) Hydrological characteristics of the Straits of Magellan: austral summer 1990/91 (February-March). Mem Biol Mar Oceanogr 19: 77–81

Artegiani A, Paschini E, Andueza-Calderón (1991) Physical oceanography of the Straits of Magellan. Nat Sc Com Ant, Magellan Cruise, February-March 1991, Data Rep I: 11--52

Balech E (1954) División zoogeográfica del litoral sudamericano. Rev Biol Mar Valparaíso 4: 184–195

Benassi G, Naldi M, Mckenzie KG (1992) Preliminary research on species distribution of planktonic ostracods (Halocyprididae) in the Ross Sea, Antarctica. Crustacean Biol 12: 68–78

Brattström H, Johanssen A (1983) Ecological and regional zoogeography of the marine benthic fauna of Chile. Report No 49 of the Lund University Chile Expedition 1948–1949. Sarsia 68: 283–339

Briggs JC (1974) Marine zoogeography. McGraw-Hill, New York, 475 pp

Brinton E (1962) The distribution of Pacific euphausiids. Bull Scripps Inst Oceanogr 8: 51–270

Clement A, Guzmán L (1989) Red tides in Chilean fjords. In: Okaichi T, Anderson DM, Nemoto T (eds) Red tides: biology, environmental science and toxicology. Elsevier, New York, pp 121–124

Eckman S (1953) Zoogeography of the sea. Sidgwick and Jackson, London, 417 pp

Guglielmo L, Antezana T, Costanzo G, Zagami G (1991) Zooplankton communities in the Straits of Magellan. Mem Biol Mar Oceanogr 19: 157–161

Guglielmo L, Arena G, Costanzo G, Crescenti N, Vanucci S, Zagami G (1993) Zooplankton ecology in the Straits of Magellan. Nat Sc Com Ant, Magellan Cruise, Feb–March 1991, data report II, pp. 137–250

Hedgpeth JW (1969) Introduction to Antarctic zoogeography. Antarctic Map Folio Ser 11: 1–9, 15 figs

Johnson MV, Brinton E (1963) Biological species, water masses and currents. In: Hill MN (ed) The sea. Interscience Publ, New York, pp 381–414

Knox GA (1960) Littoral ecology and biogeography of the southern ocean. Proc R Soc Lond, Ser B 152: 577–624

Marín V, Antezana T (1985) Species composition and relative abundance of copepods in Chilean fjords. J Plankton Res 7: 961–966

Mazzocchi MG, Zagami G, Ianora A, Guglielmo L, Crescenti N, Hure J (1995) Atlas of marine zooplankton, Straits of Magellan, vol 1. In: Guglielmo L, Ianora A (eds) Copepods. Springer, Berlin Heidelberg New York, 279 pp, 183 figs

Pickard GL (1973) Water structure in Chilean fjords. In: Fraser R (ed) Oceanography of the South Pacific 1972. NZ Nat Comm UNESCO, Wellington, pp 95–104

Schilder FA (1956) Lehrbuch der allgemeinen Zoogeographie. G Fischer, Jena, 150 pp

## 2.1 Sampling Techniques

Zooplankton samples were collected in the late austral summer of 1991 (from February 20 to March 4) by the R/V *Cariboo*. Nineteen stations were sampled from 30 to 1120 m depth along the main latitudinal axis of the Straits (52°–54°S latitude; 69–73°W longitude); Fig. 2.1. Details of station and sample data are given in Table 2.1 of the first volume of this Atlas, whereas station and sample data for each taxonomic group are given in Tables 2.1 and 2.2 of this volume.

Samples were taken with an EZNET BIONESS (Easy Zooplankton Net – Bedford Institute of Oceanography Net and Environmental Sensing Sytem) equipped with ten opening/closing nets with a mouth opening of 0.25 m². The EZNET BIONESS was fitted with three types of plankton nets; one 1000-μm net followed by five 500-μm and four 230-μm nets alternatively arranged on the main frame. The data from both nets are reported in the present study.

The EZNET BIONESS continuously recorded temperature, salinity and light attenuance with a KMS II (ME Meerestichnik Elektronic GmbH) multiparametric probe. Oxygen concentration was determined with a calibrated oxygen electrode mounted on the KMS II. Flow velocity through the nets was monitored with external and internal acoustic doppler flowmeters (SM 21H-ME Meerestichnik Elektonic GmbH). Chlorophyll a concentration was measured with a Mie-Backscat Fluorometer (Haardt Optik Mikroelektronik) mounted on the EZNET BIONESS. Details of the flow speed and sample depths are given in Chapter 2 of the first volume.

## 2.2 Identification Cards

The presentation of the species is organized in five chapters, one for each taxonomic group. Each chapter gives a systematic account of the group and a brief synonymy of the species. When possible, female and male specimens are described, reporting only morphological characters useful for a rapid and precise species identification. Line drawings are presented of those anatomical parts which are essential to correctly identify the species. SEM micrographs are intended as companion illustrations to line drawings, revealing many anatomical details not always evident in the light microscope.

The data on species spatial distributions are expressed as numbers $100^{-3}$. All data are plotted in distribution maps with no consideration for the depths at which different samples were collected. The vertical distribution of each species, given below the distribution maps, was obtained by grouping data for three geographic areas referred to in the text as:

Pacific Area, from the Pacific entrance to Carlos III Island (Sts. 5, 6, 7, 9, 10, 11);

Central Area, from Carlos III Island to Pelicano Point (Sts. 12, 13, 14, 15, 16, 17, 18, 19);

Atlantic Area, from Pelicano Point to the Atlantic entrance (Sts. 20, 21, 22, 23, 26).

Temperature (T) and fluorescence (F) profiles are shown for a key station within each area: St. 9 for the Pacific area, St. 15 for the Central area and St. 20 for the Atlantic area. The conventional unit (F) is used for in vivo fluorescence in the range 0–10 V, corresponding to 0–50 mg/m³ chl a, with a resolution of 0.1 mg/m³ and an accuracy variability of less than 10 %.

Each chapter ends with the bibliography relevant to each taxonomic group and an appendix with species abundance data for the stations and depths sampled during the cruise.

## 2.3  Line Drawings and Scanning Electron Micrographs

Original ink figures, obtained by use of a camera lucida, were processed by a computer graphic system (Apple MacIntosh Quadra with 20 MB RAM and 128 MB magneto-optic disc). The images were scanned through a UMAX 600 DPI scanner and then proportionately reduced and paged through an Aldus Free Hand drawing program.

Specimens for scanning electron microscopy were prepared according to the method given in the first volume on copepods. Preserved samples were washed and left in distilled water for at least 24 h. Amphipods, euphausiids and mysids were then postfixed in a 2 % $OsO_4$ solution for 1 h prior to dehydration in a graded alcohol series: 50 % for 30 min, 70 % for 30 min, 90 % for 30 min, followed by three washings in 100 % alcohol for 30 min each. The procedure was the same for ostracods except that they were postfixed in a 1 % $OsO_4$ solution and then dehydrated. For chaetognaths, specimens were dehydrated only in the above-mentioned graded alcohol series. All specimens were then critical point-dried with liquid $CO_2$ for 3–5 h. $CO_2$ was changed every hour. Specimens were then slowly brought to a temperature of 50 °C and a pressure of 99 atm (1450 psi). The pressure was gradually brought back to normal within 30 min and specimens were slowly cooled to ambient temperature. The specimens were then mounted on stubs with adhesive tape and coated with gold for 3 min at a setting of 2.2 kV and 10 mA of the sputtering apparatus.

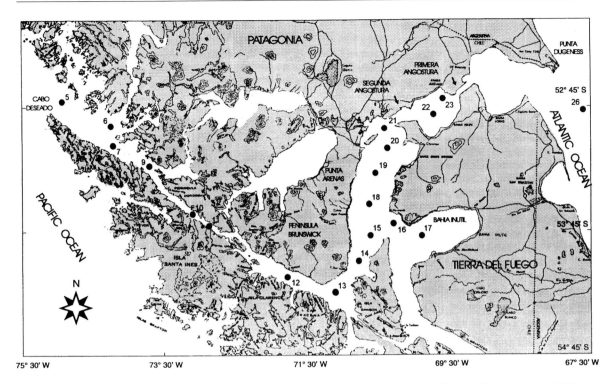

**Fig. 2.1.** Zooplankton sampling sites during the R/V *Cariboo* cruise in the Straits of Magellan in February–March 1991

**Table 2.1.** Station and sample data for ostracods sampled during the R/V *Cariboo* cruise in the Straits of Magellan in February-March 1991

| Station no. | Deth sampled (m) | Volume filtered (m$^3$) | Ostracods (Ind m$^{-3}$) | Temperature (°C) | Salinity (psu) | Fluorescence (U.F.) |
|---|---|---|---|---|---|---|
| 5 | 160-140 | 53.10 | 0.97 | 10.350 | 33.008 | 0.074 |
|   | 115-100 | 32.60 | 1.61 | 10.559 | 32.931 | 0.070 |
|   | 80-60 | 45.10 | 1.64 | 9.184 | 33.008 | 0.065 |
|   | 40-20 | 32.20 | 0.71 | 9.629 | 30.193 | 0.166 |
| 6 | 455-400 | 120.80 | 1.43 | 8.163 | 33.260 | 0.046 |
|   | 300-250 | 45.10 | 6.39 | 9.069 | 33.251 | 0.050 |
|   | 200-150 | 45.90 | 17.58 | 9.808 | 33.213 | 0.054 |
|   | 100-50 | 99.10 | 6.46 | 10.090 | 31.203 | 0.124 |
| 7 | 500-400 | 79.80 | 4.02 | 8.204 | 33.256 | 0.043 |
|   | 300-250 | 56.80 | 5.53 | 8.446 | 33.192 | 0.045 |
|   | 200-150 | 63.70 | 1.54 | 9.486 | 33.113 | 0.052 |
|   | 100-50 | 58.20 | 11.47 | 9.235 | 31.339 | 0.087 |
| 9 | 700-600 | 79.00 | 0.31 | 8.204 | 33.252 | 0.043 |
|   | 400-300 | 62.30 | 4.33 | 8.393 | 33.236 | 0.046 |
|   | 200-140 | 23.70 | 10.61 | 9.064 | 32.975 | 0.052 |
|   | 100-50 | 44.60 | 7.73 | 8.720 | 31.493 | 0.092 |
| 11 | 500-400 | 191.00 | 1.38 | 8.244 | 33.242 | 0.043 |
|   | 300-250 | 79.10 | 5.68 | 8.421 | 33.132 | 0.045 |
|   | 200-150 | 73.00 | 11.30 | 8.596 | 32.803 | 0.052 |
|   | 100-50 | 36.90 | 2.66 | 8.901 | 30.479 | 0.187 |
|   | 400-300 | 302.70 | 5.35 | 8.272 | 33.222 | 0.045 |
|   | 250-200 | 130.40 | 9.60 | 8.375 | 33.102 | 0.050 |
|   | 150-100 | 94.70 | 10.06 | 8.526 | 32.557 | 0.067 |
|   | 50-25 | 32.80 | 12.88 | 9.342 | 29.094 | 0.263 |
| 12 | 130-100 | 20.90 | 14.33 | 7.738 | 30.763 | 0.084 |
|   | 100-50 | 49.60 | 2.70 | 7.998 | 30.605 | 0.150 |
|   | 50-25 | 47.00 | 0.48 | 8.223 | 30.448 | 0.287 |
|   | 25-0 | 38.40 | 0.30 | 8.464 | 30.183 | 0.210 |
| 13 | 400-300 | 249.80 | 18.26 | 6.926 | 31.112 | 0.053 |
|   | 250-200 | 48.80 | 111.76 | 7.225 | 31.062 | 0.058 |
|   | 150-100 | 68.70 | 27.93 | 7.765 | 30.812 | 0.066 |
|   | 50-25 | 38.60 | 5.48 | 8.385 | 30.382 | 0.233 |
| 14 | 400-300 | 156.30 | 167.90 | 6.883 | 31.082 | 0.052 |
|   | 250-200 | 65.90 | 202.23 | 7.406 | 30.989 | 0.054 |
|   | 150-100 | 85.00 | 70.63 | 7.447 | 30.780 | 0.077 |
|   | 90-50 | 56.90 | 6.25 | 8.332 | 30.525 | 0.114 |
| 15 | 300-200 | 155.40 | 91.20 | 6.922 | 31.004 | 0.054 |
|   | 150-100 | 36.00 | 76.60 | 7.614 | 30.695 | 0.066 |
|   | 80-60 | 26.00 | 126.59 | 8.258 | 30.537 | 0.093 |
|   | 40-20 | 26.10 | 426.01 | 8.594 | 30.358 | 0.194 |
| 16 | 80-60 | 30.70 | 40.52 | 7.619 | 30.617 | 0.071 |
|   | 60-40 | 55.80 | 7.58 | 8.590 | 30.482 | 0.215 |
|   | 40-20 | 22.40 | 3.56 | 8.946 | 30.425 | 0.248 |
|   | 20-0 | 36.10 | 0.69 | 9.548 | 30.199 | 0.526 |
| 17 | 120-100 | 41.40 | 632.21 | 7.021 | 30.647 | 0.063 |
|   | 80-60 | 43.90 | 18.30 | 7.836 | 30.597 | 0.083 |
|   | 40-30 | 20.00 | 9.49 | 9.003 | 30.406 | 0.291 |
|   | 20-10 | 18.70 | 11.17 | 9.298 | 30.341 | 0.530 |
| 18 | 160-120 | 92.00 | 436.83 | 7.037 | 30.914 | 0.053 |
|   | 100-80 | 43.40 | 21.17 | 7.225 | 30.804 | 0.055 |
|   | 60-40 | 39.20 | 11.11 | 8.221 | 30.612 | 0.128 |
|   | 20-10 | 14.50 | 14.42 | 9.237 | 30.453 | 0.548 |
|   | 20-0 | 23.20 | 0.15 | 9.079 | 30.380 | 0.417 |

**Table 2.1.** (continued)

| Station no. | Deth sampled (m) | Volume filtered (m$^3$) | Ostracods (Ind m$^{-3}$) | Temperature (°C) | Salinity (psu) | Fluorescence (U.F.) |
|---|---|---|---|---|---|---|
| 19 | 100-75 | 57.70 | 27.14 | 8.074 | 30.635 | 0.093 |
|    | 75-50  | 46.30 | 10.44 | 8.534 | 30.552 | 0.115 |
|    | 50-25  | 42.30 | 4.43  | 8.834 | 30.500 | 0.211 |
|    | 25-0   | 29.70 | 0.69  | 8.996 | 30.403 | 0.309 |
| 20 | 80-75  | 52.10 | 13.07 | 8.699 | 30.525 | 0.186 |
|    | 60-40  | 35.60 | 3.08  | 8.836 | 30.495 | 0.219 |
|    | 40-20  | 20.30 | 2.81  | 8.988 | 30.487 | 0.256 |
| 21 | 40-20  | 60.95 | 3.31  | 9.169 | 30.522 | 0.228 |
|    | 20-0   | 39.20 | 1.66  | 9.121 | 30.517 | 0.250 |
| 22 | 30-20  | 17.20 | 0.40  | 9.805 | 30.542 | 0.233 |
|    | 20-10  | 51.70 | 0.37  | 9.788 | 30.543 | 0.233 |
|    | 10-0   | 24.10 | 0.14  | 9.761 | 30.539 | 0.236 |
| 23 | 30-0   | 46.70 | 0.02  | 10.217 | 30.754 | 0.330 |
| 26 | 30-20  | 22.15 | 0.25  | 9.299 | 32.654 | 0.241 |
|    | 20-0   | 28.40 | 0.04  | 9.303 | 32.649 | 0.234 |

**Table 2.2.** Station and sample data for amphipods, mysids, euphausiids and chaetognaths sampled during the R/V *Cariboo* cruise in the Straits of Magellan

| Station no. | Depth sampled (m) | Volume filtered (m$^3$) | Amphipods (Ind m$^{-3}$) | Mysids (Ind $^{-3}$) | Euphausiids (Ind m$^{-3}$) | Chaetognaths (Ind m$^{-3}$) | Temperature (°C) | Salinity (psu) | Fluorescence (U.F.) |
|---|---|---|---|---|---|---|---|---|---|
| 5 | 200-160 | 121.80 | 0.30 | 0.00 | 1.70 | 14.00 | 9.629 | 33.197 | 0.065 |
| | 160-140 | 53.10 | 0.04 | 0.00 | 0.56 | 6.22 | 10.350 | 33.008 | 0.074 |
| | 140-115 | 60.80 | 0.02 | 0.00 | 0.21 | 0.68 | 10.442 | 33.007 | 0.068 |
| | 115-100 | 32.60 | 0.11 | 0.00 | 0.39 | 3.75 | 10.559 | 32.931 | 0.070 |
| | 100-80 | 59.30 | 0.04 | 0.00 | 0.10 | 0.42 | 10.207 | 32.622 | 0.067 |
| | 80-60 | 45.10 | 0.08 | 0.00 | 0.00 | 3.29 | 9.184 | 33.008 | 0.065 |
| | 60-40 | 35.90 | 0.03 | 0.00 | 0.70 | 0.22 | 8.909 | 31.025 | 0.092 |
| | 40-20 | 32.20 | 0.00 | 0.00 | 1.35 | 0.25 | 9.629 | 30.193 | 0.166 |
| | 20-0 | 41.50 | 0.11 | 0.00 | 0.41 | 0.36 | 9.914 | 29.961 | 0.137 |
| 6 | 500-450 | 76.30 | 0.02 | 0.75 | 0.00 | 0.74 | 8.155 | 33.267 | 0.046 |
| | 450-400 | 120.80 | 0.01 | 0.39 | 0.00 | 0.88 | 8.163 | 33.260 | 0.046 |
| | 400-300 | 220.80 | 0.02 | 0.13 | 0.00 | 0.34 | 8.416 | 33.233 | 0.047 |
| | 300-250 | 45.10 | 0.05 | 0.30 | 0.00 | 0.48 | 9.069 | 33.251 | 0.050 |
| | 250-200 | 53.10 | 0.07 | 0.04 | 0.04 | 0.38 | 9.688 | 33.284 | 0.053 |
| | 200-150 | 45.90 | 0.27 | 0.00 | 1.47 | 1.10 | 9.808 | 33.213 | 0.054 |
| | 150-100 | 47.30 | 0.07 | 0.00 | 0.46 | 1.47 | 9.977 | 33.002 | 0.056 |
| | 100-50 | 99.10 | 0.06 | 0.00 | 0.81 | 1.02 | 10.090 | 31.203 | 0.124 |
| | 50-0 | 124.10 | 0.55 | 0.00 | 0.00 | 0.11 | 10.109 | 29.425 | 0.212 |
| 7 | 600-500 | 80.20 | 0.03 | 0.13 | 0.01 | 0.71 | 8.107 | 33.269 | 0.043 |
| | 500-400 | 79.80 | 0.03 | 0.49 | 0.00 | 0.12 | 8.204 | 33.256 | 0.043 |
| | 400-300 | 93.60 | 0.38 | 0.24 | 0.00 | 0.07 | 8.278 | 33.227 | 0.043 |
| | 300-250 | 56.80 | 0.02 | 0.66 | 0.02 | 0.44 | 8.446 | 33.192 | 0.045 |
| | 250-200 | 88.10 | 0.08 | 0.27 | 0.00 | 0.51 | 8.946 | 33.194 | 0.048 |
| | 200-150 | 63.70 | 0.13 | 0.02 | 4.79 | 1.63 | 9.486 | 33.113 | 0.052 |
| | 150-100 | 28.50 | 0.04 | 0.00 | 1.13 | 1.77 | 9.604 | 32.789 | 0.049 |
| | 100-50 | 58.20 | 0.12 | 0.00 | 0.31 | 0.14 | 9.235 | 31.339 | 0.087 |
| | 50-0 | 73.50 | 0.56 | 0.02 | 0.40 | 0.81 | 9.498 | 29.717 | 0.177 |
| 9 | 900-700 | 116.30 | 0.03 | 0.04 | 0.02 | 0.32 | 8.208 | 33.252 | 0.046 |
| | 700-600 | 79.00 | 0.03 | 0.00 | 0.00 | 0.20 | 8.204 | 33.252 | 0.043 |
| | 600-400 | 260.60 | 0.01 | 0.07 | 0.02 | 0.07 | 8.297 | 33.255 | 0.044 |
| | 400-300 | 62.30 | 0.00 | 0.06 | 0.00 | 0.0E | 8.393 | 33.236 | 0.046 |
| | 300-200 | 76.40 | 0.00 | 0.18 | 0.00 | 0.30 | 8.593 | 33.126 | 0.045 |
| | 200-140 | 23.70 | 0.19 | 0.00 | 9.89 | 3.72 | 9.064 | 32.975 | 0.052 |
| | 140-100 | 25.10 | 0.00 | 0.05 | 1.73 | 3.23 | 9.101 | 32.188 | 0.053 |
| | 100-50 | 44.60 | 0.08 | 0.00 | 2.23 | 2.38 | 8.720 | 31.493 | 0.092 |
| | 50-0 | 55.00 | 0.13 | 0.00 | 0.33 | 0.31 | 9.224 | 26.350 | 0.151 |
| 10 | 600-500 | 133.70 | 0.00 | 0.00 | 0.00 | 0.24 | 8.210 | 33.250 | 0.042 |
| | 500-400 | 191.00 | 0.00 | 0.01 | 0.00 | 0.34 | 8.244 | 33.242 | 0.043 |
| | 400-300 | 169.30 | 0.00 | 0.00 | 0.00 | 0.20 | 8.322 | 33.219 | 0.043 |
| | 300-250 | 79.10 | 0.01 | 0.01 | 0.04 | 0.35 | 8.421 | 33.132 | 0.045 |
| | 250-200 | 88.90 | 0.00 | 0.01 | 0.00 | 0.36 | 8.515 | 32.976 | 0.049 |
| | 200-150 | 73.00 | 0.00 | 0.08 | 0.03 | 1.74 | 8.596 | 32.803 | 0.052 |
| | 150-100 | 24.90 | 0.51 | 0.05 | 0.09 | 1.65 | 8.503 | 32.325 | 0.067 |
| | 100-50 | 36.90 | 1.39 | 0.00 | 0.00 | 6.13 | 8.901 | 30.479 | 0.187 |
| | 50-0 | 69.10 | 0.58 | 0.00 | 0.03 | 0.05 | 9.600 | 28.395 | 0.231 |
| 11 | 500-400 | 206.10 | 0.02 | 0.11 | 0.01 | 1.09 | 8.250 | 33.248 | 0.045 |
| | 400-300 | 302.70 | 0.02 | 0.04 | 0.01 | 0.84 | 8.272 | 33.222 | 0.045 |
| | 300-250 | 154.90 | 0.00 | 0.00 | 0.01 | 0.19 | 8.307 | 33.189 | 0.048 |
| | 250-200 | 130.40 | 0.00 | 0.01 | 0.04 | 0.32 | 8.375 | 33.102 | 0.050 |
| | 200-150 | 120.90 | 0.04 | 0.00 | 0.09 | 0.67 | 8.475 | 32.970 | 0.052 |
| | 150-100 | 94.70 | 0.10 | 0.01 | 0.12 | 1.32 | 8.526 | 32.557 | 0.067 |
| | 100-50 | 87.10 | 0.30 | 0.03 | 0.60 | 1.82 | 8.760 | 31.177 | 0.144 |
| | 50-25 | 32.80 | 1.43 | 0.04 | 3.31 | 0.70 | 9.342 | 29.094 | 0.263 |
| | 25-0 | 36.50 | 2.98 | 0.00 | 3.32 | 1.38 | 9.747 | 27.920 | 0.198 |

**Table 2.2.** (continued)

| Station no. | Depth sampled (m) | Volume filtered (m$^3$) | Amphipods (Ind m$^{-3}$) | Mysids (Ind m$^{-3}$) | Euphausiids (Ind m$^{-3}$) | Chaetognaths (Ind m$^{-3}$) | Temperature (°C) | Salinity (psu) | Fluorescence (U.F.) |
|---|---|---|---|---|---|---|---|---|---|
| 12 | 130-100 | 29.50 | 0.29 | 0.00 | 2.52 | 0.03 | 7.674 | 30.788 | 0.079 |
|    | 100-50 | 54.20 | 0.25 | 0.00 | 0.01 | 0.04 | 7.956 | 30.662 | 0.134 |
|    | 50-25 | 38.50 | 1.09 | 0.00 | 0.03 | 0.00 | 8.258 | 30.416 | 0.214 |
|    | 25-0 | 49.50 | 1.86 | 0.00 | 0.03 | 0.00 | 8.483 | 27.538 | 0.172 |
| 13 | 450-400 | 93.80 | 0.01 | 0.66 | 0.00 | 0.15 | 6.903 | 31.122 | 0.052 |
|    | 400-300 | 249.80 | 0.01 | 0.68 | 0.01 | 0.15 | 6.926 | 31.112 | 0.053 |
|    | 300-250 | 59.90 | 0.00 | 0.31 | 0.00 | 0.11 | 6.981 | 31.088 | 0.055 |
|    | 250-200 | 48.80 | 0.28 | 1.03 | 0.00 | 0.26 | 7.225 | 31.062 | 0.058 |
|    | 200-150 | 52.70 | 0.33 | 0.59 | 0.13 | 2.80 | 7.601 | 30.982 | 0.060 |
|    | 150-100 | 68.70 | 0.12 | 0.00 | 0.85 | 1.13 | 7.765 | 30.812 | 0.066 |
|    | 100-50 | 88.50 | 0.00 | 0.00 | 0.76 | 0.04 | 8.058 | 30.552 | 0.132 |
|    | 50-25 | 38.60 | 1.36 | 0.09 | 0.18 | 0.30 | 8.385 | 30.382 | 0.233 |
|    | 25-0 | 34.10 | 2.88 | 0.03 | 0.00 | 0.00 | 8.440 | 30.371 | 0.250 |
| 14 | 500-400 | 149.70 | 0.07 | 0.35 | 0.02 | 0.21 | 6.857 | 31.130 | 0.053 |
|    | 400-300 | 156.30 | 0.00 | 0.32 | 0.02 | 0.53 | 6.883 | 31.082 | 0.052 |
|    | 300-250 | 101.90 | 0.01 | 0.54 | 0.12 | 0.84 | 6.895 | 31.045 | 0.054 |
|    | 250-200 | 65.90 | 0.07 | 0.76 | 3.10 | 1.25 | 7.406 | 30.989 | 0.054 |
|    | 200-150 | 78.60 | 0.36 | 0.06 | 21.26 | 6.03 | 7.228 | 30.865 | 0.062 |
|    | 150-100 | 85.00 | 0.27 | 0.00 | 10.08 | 2.26 | 7.447 | 30.780 | 0.077 |
|    | 100-90 | 17.00 | 0.00 | 0.00 | 0.87 | 0.13 | 8.224 | 30.546 | 0.095 |
|    | 90-50 | 56.90 | 0.04 | 0.00 | 0.16 | 0.10 | 8.332 | 30.525 | 0.114 |
|    | 50-0 | 44.00 | 0.26 | 0.00 | 21.58 | 0.10 | 8.667 | 30.450 | 0.271 |
| 15 | 434-300 | 225.80 | 0.01 | 0.69 | 0.00 | 0.32 | 6.798 | 31.143 | 0.054 |
|    | 300-200 | 155.40 | 0.08 | 0.34 | 0.01 | 0.32 | 6.922 | 31.004 | 0.054 |
|    | 200-150 | 54.60 | 0.06 | 0.44 | 0.04 | 0.25 | 7.165 | 30.842 | 0.057 |
|    | 150-100 | 36.00 | 0.16 | 0.13 | 0.03 | 0.16 | 7.614 | 30.695 | 0.066 |
|    | 100-80 | 15.30 | 0.08 | 0.22 | 0.08 | 0.00 | 8.067 | 30.579 | 0.086 |
|    | 80-60 | 26.00 | 0.40 | 0.04 | 0.04 | 0.53 | 8.258 | 30.537 | 0.093 |
|    | 60-40 | 26.20 | 0.17 | 0.09 | 0.17 | 0.09 | 8.420 | 30.481 | 0.106 |
|    | 40-20 | 26.10 | 0.61 | 0.04 | 3.85 | 1.58 | 8.594 | 30.358 | 0.194 |
|    | 20-0 | 29.80 | 0.27 | 0.00 | 29.49 | 5.83 | 8.672 | 30.244 | 0.234 |
| 16 | 80-60 | 45.40 | 0.37 | 0.00 | 0.76 | 0.48 | 7.554 | 30.622 | 0.067 |
|    | 60-40 | 50.10 | 0.45 | 0.00 | 0.02 | 0.00 | 8.406 | 30.514 | 0.161 |
|    | 40-20 | 20.40 | 1.61 | 0.12 | 0.29 | 0.46 | 9.185 | 30.362 | 0.328 |
|    | 20-0 | 29.90 | 1.45 | 0.00 | 4.75 | 0.00 | 9.501 | 30.242 | 0.362 |
| 17 | 140-120 | 31.30 | 0.33 | 0.00 | 2.41 | 15.52 | 6.770 | 30.653 | 0.061 |
|    | 120-100 | 41.40 | 0.03 | 0.00 | 2.71 | 15.50 | 7.021 | 30.647 | 0.063 |
|    | 100-80 | 43.80 | 0.13 | 0.00 | 0.19 | 0.08 | 7.421 | 30.629 | 0.071 |
|    | 80-60 | 43.90 | 1.20 | 0.00 | 0.08 | 0.18 | 7.836 | 30.597 | 0.083 |
|    | 60-40 | 41.70 | 0.85 | 0.00 | 0.00 | 0.00 | 8.405 | 30.509 | 0.152 |
|    | 40-30 | 20.00 | 2.40 | 0.00 | 0.06 | 0.00 | 9.003 | 30.406 | 0.291 |
|    | 30-20 | 20.70 | 1.66 | 0.00 | 0.72 | 0.06 | 9.236 | 30.363 | 0.422 |
|    | 20-10 | 18.70 | 0.98 | 0.00 | 0.00 | 0.18 | 9.298 | 30.341 | 0.506 |
|    | 10-0 | 21.20 | 0.05 | 0.00 | 0.49 | 0.27 | 9.312 | 30.330 | 0.560 |
| 18 | 180-160 | 46.90 | 0.41 | 0.02 | 14.38 | 18.52 | 6.976 | 30.961 | 0.053 |
|    | 160-120 | 92.00 | 0.55 | 0.00 | 5.06 | 5.21 | 7.037 | 30.914 | 0.053 |
|    | 120-100 | 43.10 | 0.03 | 0.00 | 0.48 | 0.08 | 7.121 | 30.852 | 0.053 |
|    | 100-80 | 43.40 | 0.21 | 0.00 | 0.24 | 0.21 | 7.225 | 30.804 | 0.055 |
|    | 80-60 | 39.40 | 0.00 | 0.00 | 0.06 | 0.00 | 7.419 | 30.744 | 0.056 |
|    | 60-40 | 39.20 | 0.18 | 0.00 | 0.09 | 0.06 | 8.221 | 30.612 | 0.128 |
|    | 40-20 | 36.90 | 0.09 | 0.00 | 0.06 | 0.03 | 9.047 | 30.486 | 0.359 |
|    | 20-10 | 14.50 | 0.71 | 0.00 | 0.55 | 0.08 | 9.237 | 30.453 | 0.548 |
|    | 10-0 | 17.60 | 0.26 | 0.00 | 0.91 | 0.58 | 9.309 | 30.444 | 0.488 |

**Table 2.2.** (continued)

| Station no. | Depth sampled (m) | Volume filtered (m³) | Amphipods (Ind m⁻³) | Mysids (Ind m⁻³) | Euphausiids (Ind m⁻³) | Chaetognaths (Ind m⁻³) | Temperature (°C) | Salinity (psu) | Fluorescence (U.F.) |
|---|---|---|---|---|---|---|---|---|---|
| 19 | 106-75 | 54.60 | 0.32 | 0.02 | 0.06 | 1.51 | 8.045 | 30.632 | 0.078 |
|    | 75-50  | 45.30 | 0.46 | 0.00 | 0.46 | 0.53 | 8.538 | 30.558 | 0.122 |
|    | 50-25  | 50.60 | 0.68 | 0.00 | 0.89 | 0.17 | 8.792 | 30.492 | 0.182 |
|    | 25-0   | 35.10 | 0.22 | 0.00 | 2.38 | 0.47 | 9.005 | 26.858 | 0.241 |
| 20 | 80-75  | 39.10 | 0.35 | 0.00 | 0.11 | 0.90 | 8.768 | 30.510 | 0.199 |
|    | 75-60  | 53.10 | 0.18 | 0.00 | 0.17 | 0.09 | 8.831 | 30.499 | 0.216 |
|    | 60-40  | 35.60 | 0.26 | 0.00 | 0.23 | 0.16 | 8.836 | 30.495 | 0.219 |
|    | 40-20  | 29.90 | 1.94 | 0.00 | 0.05 | 0.07 | 8.981 | 30.468 | 0.268 |
|    | 20-0   | 23.40 | 1.08 | 0.00 | 0.03 | 0.00 | 9.045 | 30.424 | 0.379 |
| 21 | 40-20  | 55.60 | 1.86 | 0.00 | 0.08 | 0.05 | 9.144 | 30.520 | 0.226 |
|    | 20-0   | 53.40 | 2.11 | 0.00 | 0.02 | 0.07 | 9.165 | 30.519 | 0.212 |
| 22 | 30-20  | 35.00 | 0.43 | 0.00 | 0.03 | 0.00 | 9.800 | 30.548 | 0.232 |
|    | 20-10  | 23.10 | 0.49 | 0.00 | 0.00 | 0.00 | 9.787 | 30.544 | 0.230 |
|    | 10-0   | 24.80 | 0.17 | 0.00 | 0.00 | 0.00 | 9.777 | 30.568 | 0.192 |
| 23 | 30-20  | 31.50 | 1.23 | 0.00 | 0.00 | 0.00 | 10.210 | 30.756 | 0.242 |
|    | 20-10  | 31.20 | 2.10 | 0.00 | 0.00 | 0.00 | 10.213 | 30.755 | 0.245 |
|    | 10-0   | 34.20 | 0.87 | 0.00 | 0.00 | 0.00 | 10.220 | 30.755 | 0.285 |
| 26 | 30-20  | 24.30 | 11.02 | 1.39 | 0.20 | 0.12 | 9.304 | 32.651 | 0.244 |
|    | 20-10  | 21.50 | 2.08 | 0.03 | 0.00 | 0.06 | 9.296 | 32.652 | 0.228 |
|    | 10-0   | 36.60 | 2.56 | 0.05 | 0.00 | 0.04 | 9.297 | 32.651 | 0.234 |

## *Chapter 3*   **Amphipods**

G. Costanzo and N. Crescenti

# Systematic Account

## 3.1 *Cyphocaris faurei* Barnard, 1916

*Cyphocaris faurei* Barnard, 1916: p. 117, pl. 26 fig. 4 – Schellenberg, 1926: p. 215, figs. 2e, 11, 12, pl. 5 fig. 4 – Birstein and Vinogradov, 1960: p. 169 – Andres, 1983: p. 185.

Body length –
    females: 18.8–27 mm (6 specimens)

**Female.** Head a little shorter than half the first pereon segment. Eyes pear-shaped. 1st segment of pereon slightly swollen frontally but not surpassing head. 4th coxal plate obovate; anterior margin strongly convex, posterior margin concave on both sides of a small tooth. 5th coxal plate subrectangular with a distinct furrow running along inferior margin. 4th segment of pleon concave at base. 5th pereopod with 2nd article prolonged posteriorly in spiniform process that exceeds end of 5th article, posterior margin not denticulated. 2nd article of 7th pereopod with 7 small teeth along posterior margin. Telson long, narrow and cleft for 3/4 of length.

### Remarks

The shape of the P5 (as also shown by Schellenberg 1926 p. 204) leaves no doubt as to the identity of *C. faurei* specimens of the Magellan area. Of the 85 specimens sampled, only 6 were adults and all 6 were females. The only ovigerous female was 22 mm in length, as also reported by Barnard (1932) for two ovigerous females from the South and Southeast Atlantic, respectively.

### Distribution

Lowry and Bullock (1976) report the species as cosmopolitan with a depth range from 0–1200 m. Schellenberg (1926) reported a depth range from 0–2500 m; Birstein and Vinogradov (1960) a depth of 1910 m. In the Straits of Magellan, maximum densities occurred at the Pacific entrance to the Straits. Abundances diminished very rapidly towards the Atlantic, even though the species was present off Paso Largo and Cape Froward, until St. 15 off Paso Ancho. The species had a very wide vertical distribution range in both the Pacific and Central areas, with maximum densities occurring from 100–200 m.

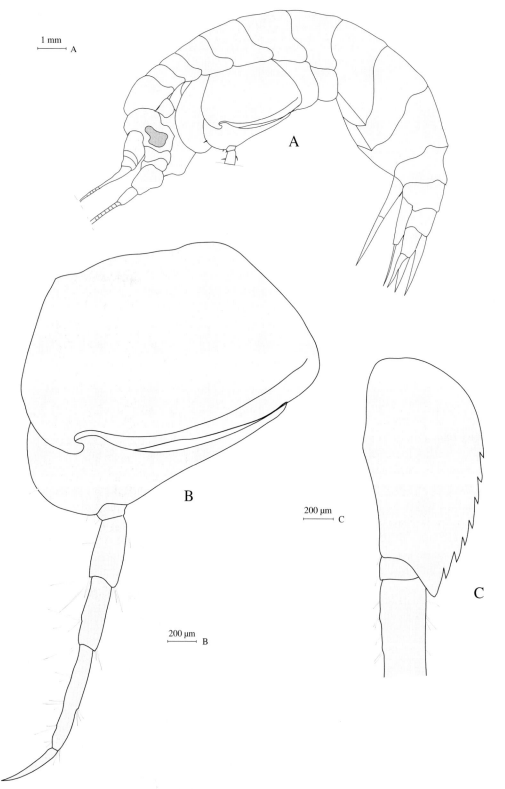

**Fig. 3.1.1A–C.** *Cyphocaris faurei*. Female: **A** whole animal lateral view (Modified after Schellenberg 1926); **B** pereopod 5; **C** pereopod 7

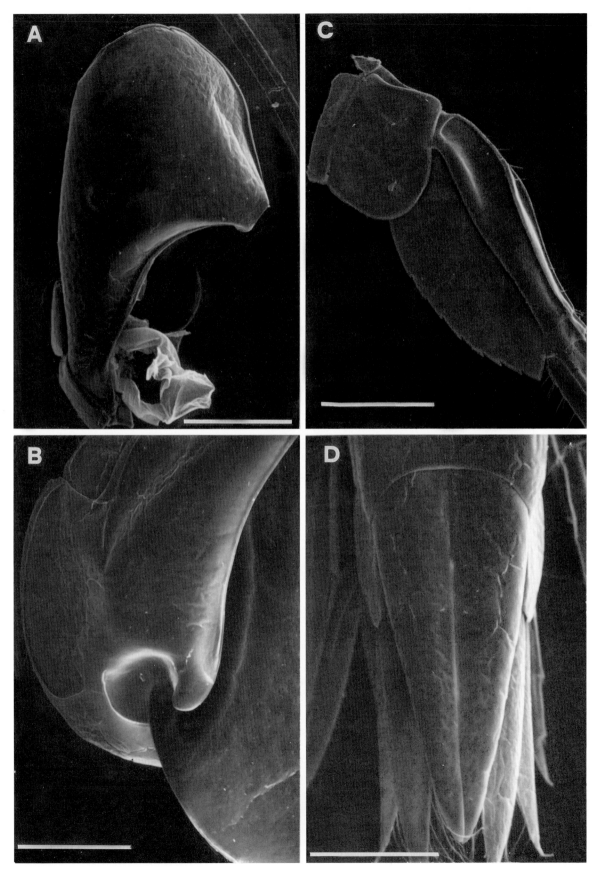

**Fig. 3.1.2A–D.** *Cyphocaris fauri*. Female: **A** 4th coxal plate; **B** pereopod 7 (part); **C** detail of 2nd article of 5th peropod; **D** telson. *Bars* **A, B** 1 mm; **C, D** 500 μm

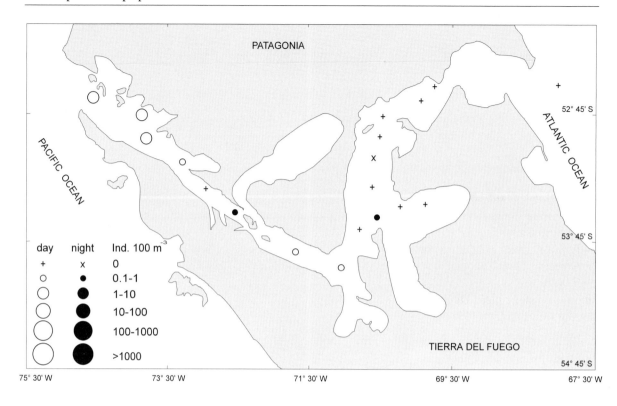

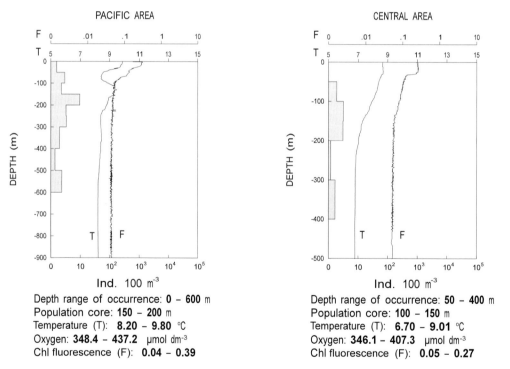

Depth range of occurrence: **0 – 600** m
Population core: **150 – 200** m
Temperature (T): **8.20 – 9.80** °C
Oxygen: **348.4 – 437.2** μmol dm⁻³
Chl fluorescence (F): **0.04 – 0.39**

Depth range of occurrence: **50 – 400** m
Population core: **100 – 150** m
Temperature (T): **6.70 – 9.01** °C
Oxygen: **346.1 – 407.3** μmol dm⁻³
Chl fluorescence (F): **0.05 – 0.27**

**Fig. 3.1.3.** *Cyphocaris fauri.* Distribution in the Straits of Magellan in February–March 1991

## 3.2 *Abyssorchomene plebs* Hurley, 1965

*Orchomenella plebs* Hurley, 1965: p. 109, figs. 1, 2 – Emison, 1968: p. 202, fig. 10 tabs. 10 – 12.
*Orchomene plebs* Thurston, 1974: p. 59 – Andres, 1983: p. 203.
*Abyssorchomene plebs* De Broyer, 1983: pp. 146 – 149, fig. 12a; Andres, 1990: p. 135, fig. 267.

Body length –
   males: 6.6–12 mm (4 specimens)

**Male.** Eyes kidney-shaped, almost meeting in midline. A1 with accessory flagellum of 8 articles, the first of which ending at 1st segment of primary flagellum. All articles are calceoliferous. A2 much longer than A1, with articles of primary flagellum that are mainly calceoliferous. 1st segment of urosome notched anteriorly and posteriorly bearing a distinct, rounded hump. Coxal plate of 1st gnathopod characteristically narrower proximally and wider distally, resembling an axe. 2nd gnathopod with very small dactyl, almost hidden by the setules. 2nd uropod not notched. 3rd uropod with shorter internal than external ramus; both rami bearing numerous plumose setules along inner margin. 3rd epimeral plate rounded antero-distally and almost forming right angle postero-distally.

### Remarks

According to Hurley (1965), the peculiar axe-shaped form of the coxal plate of the 1st gnathopod allows for a rapid identification of the species. Only the congeneric species *O. rossi* (now *A. rossi*) can be confused with *O. plebs* (now *A. plebs*) but differs for the 3rd uropod. The inner ramus is distinctly shorter than the 1st segment of the outer ramus in *O. plebs* (now *A. plebs*) and both rami bear plumose setules; in *O. rossi*, (now *A. rossi*) the inner ramus reaches at least the base of the 2nd segment of the outer ramus which never bears setules.

### Distribution

*A. plebs* is a circum-Antarctic species and has frequently been sampled under the sea ice in the Ross Sea and off McMurdo Sound (Littlepage and Pearse 1962). Its vertical depth range of distribution is included from 0–1000 m. In the Straits of Magellan, the species shows a very irregular distribution, being absent from the Central areas. It was sampled only in the Primera Angostura, in the Atlantic sector, and in the area of Tamar Island and off Carlos III Island, in the Pacific sector. The species occurred in deep layers from 400–600 m in the Pacific area and at shallow depths from 20–40 m in the Atlantic area.

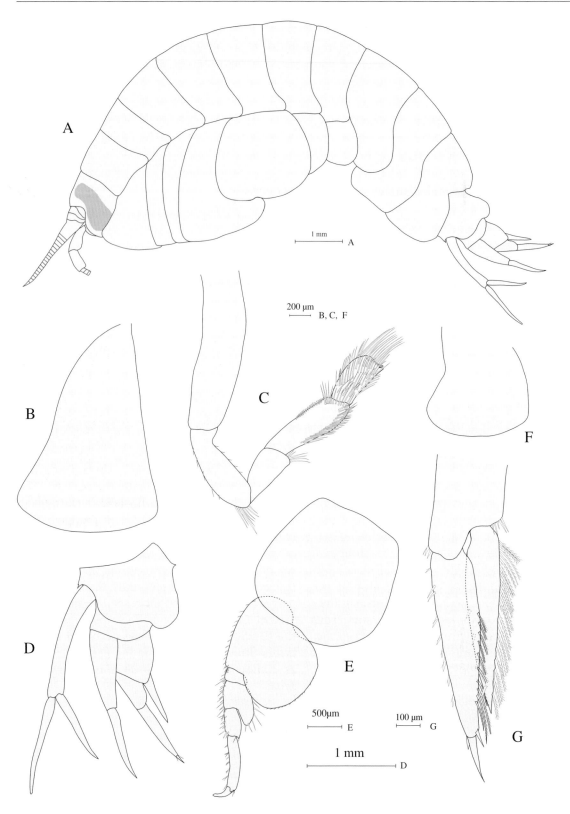

**Fig. 3.2.1A–G.** *Abyssorchomene plebs.* Male: **A** whole animal lateral view; **B** coxal plate of gnathopod 1; **C** gnathopod 2; **D** urosome lateral view; **E** pereopod 5; **F** 3rd epimeral plate; **G** uropod 3

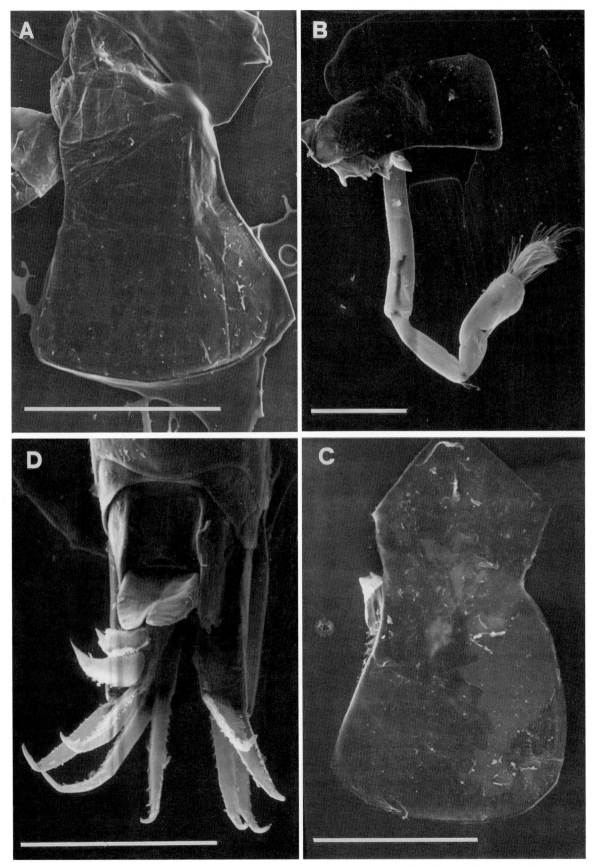

**Fig. 3.2.2A–D.** *Abyssorchomene plebs.* Female: **A** coxal plate of gnathopod 1; **B** gnathopod 2; **C** 3rd epimeral plate; **D** urosome with uropods. *Bars* **A,B,C,D** 1 mm

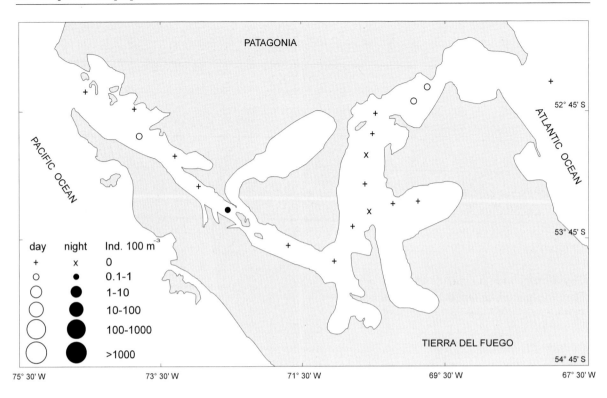

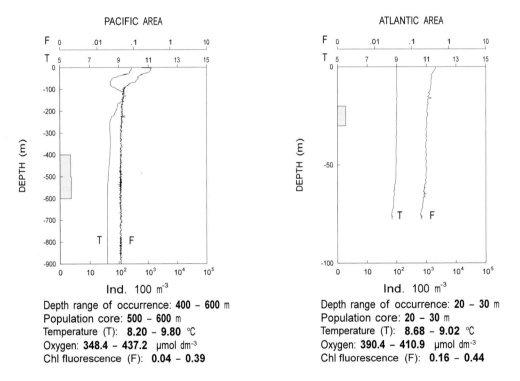

Depth range of occurrence: **400 – 600** m
Population core: **500 – 600** m
Temperature (T): **8.20 – 9.80** °C
Oxygen: **348.4 – 437.2**  µmol dm⁻³
Chl fluorescence (F): **0.04 – 0.39**

Depth range of occurrence: **20 – 30** m
Population core: **20 – 30** m
Temperature (T): **8.68 – 9.02** °C
Oxygen: **390.4 – 410.9**  µmol dm⁻³
Chl fluorescence (F): **0.16 – 0.44**

**Fig. 3.2.3.** *Abyssorchomene plebs.* Distribution in the Straits of Magellan in February–March 1991

## 3.3 *Thryphosella schellenbergi* Lowry and Bullock, 1976

*Tmetonyx serratus* Schellenberg, 1931: p. 40, fig. 19.
*Tryphosa serrata* Barnard, 1962: p. 30 (key).
*Thryphosella schellenbergi* Lowry and Bullock, 1976: pp. 7, 108 (nom. nov.).

Body length –
      female: 6.3 mm (1 specimen)

**Female.** Lateral lobes of head triangular and rounded. Eyes not visible. Posterior corner of 2nd epimeral plate bearing small tooth. Posterior border of 3rd epimeral plate convex and serrated. 1st segment of urosome with pronounced notch and with a feeble folded carina. Telson nearly twice the width and bearing 3 strong spines on both lobes. Epistome with triangular extension, rounded above upper lip. 1st maxilla with outer lobe bearing spines and inner lobe bearing 2 spinose setules.

### Remarks

*T. schellenbergi* can easily be distinguished from congeneric species by the lateral lobes of the head, which are triangular and rounded at the tip. Other characters include the epistome, the strongly depressed 1st segment of the urosome, and the posterior borders of the 2nd and 3rd epimeral plates.

### Distribution

Lowry and Bullock (1976) report the following distribution for the species: Falkland Islands, Chile, Magellanic area (Magellan Sound, Smyth Channel, Gente Grande Bay, Bahia Inutil, Ushuaia Bay). Its vertical depth range in these areas is reported to be from 0–54 m. In the present study, only a single specimen was sampled at one station in the Seconda Angostura (St. 23), where it was found from 10–30 m.

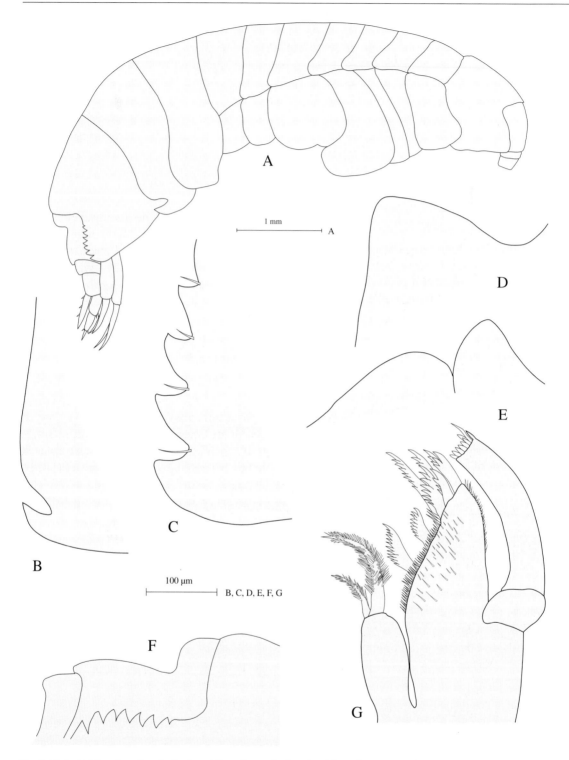

**Fig. 3.3.1 A–G.** *Thryphosella schellenbergi*. Female: **A** whole animal lateral view; **B** 2nd epimeral plate with tooth; **C** 3rd epimeral plate; **D** head lateral lobe; **E** profile of epistome and upper lip; **F** dorsal profile of pleon segments 3-5; **G** maxilla 1

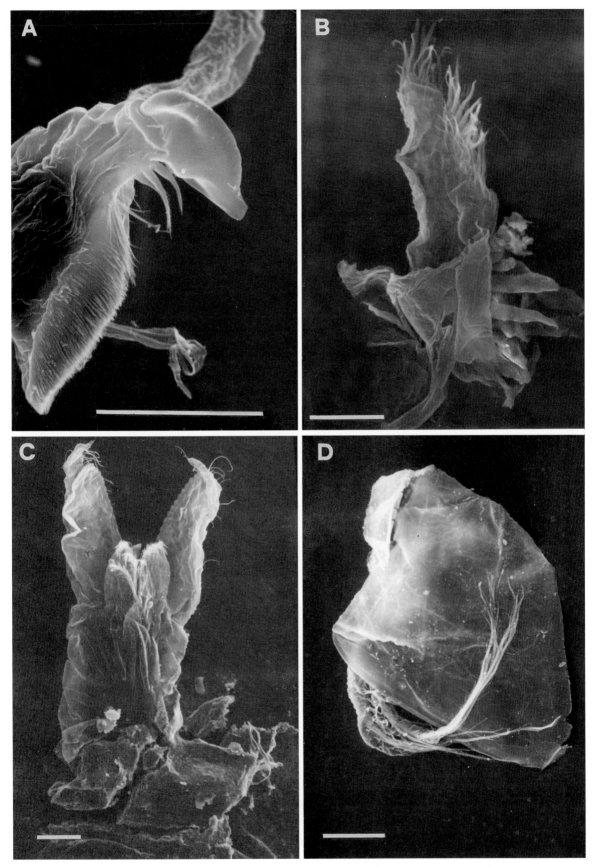

**Fig. 3.3.2A–D.** *Thryphosella schellenbergi*. Female: **A** right mandible; **B** maxilla 2; **C** maxilliped; **D** 3rd epimeral plate. *Bars* **A** 50 μm; **B,C,D** 100 μm

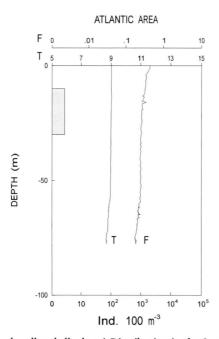

PATAGONIA

PACIFIC OCEAN

ATLANTIC OCEAN

52° 45' S

53° 45' S

54° 45' S

TIERRA DEL FUEGO

| day | night | Ind. 100 m⁻³ |
|---|---|---|
| + | x | 0 |
| ○ | • | 0.1-1 |
| ○ | ● | 1-10 |
| ○ | ● | 10-100 |
| ○ | ● | 100-1000 |
| ○ | ● | >1000 |

75° 30' W        73° 30' W        71° 30' W        69° 30' W        67° 30' W

ATLANTIC AREA

Depth range of occurrence: **10 – 30** m
Population core: **10 – 30** m
Temperature (T): **8.68 – 9.02** °C
Oxygen: **390.4 – 410.9** µmol dm⁻³
Chl fluorescence (F): **0.16 – 0.44**

**Fig. 3.3.3.** *Thryphosella schellenbergi.* Distribution in the Straits of Magellan in February–March 1991

## 3.4 *Eusirus antarcticus* Thomson, 1880

*Eusirus cuspidatus* var. *antarcticus* Thomson, 1880: p. 4
*Eusirus antarcticus* Barnard, 1961: p. 97, (key) – Bellan-Santini and Ledoyer, 1974: p. 654, pl. 8 – Thurston, 1974: p. 29.

Body length –
     juveniles: 2.8–3.7 mm (4 specimens)

**Juvenile.** 1st and 2nd segments of the peduncle of the antennulae subequal in length, bearing several teeth distally. 1st and 2nd pair of gnathopods subequal; relative proportions of articles and spinulation along palmar edge of propod of 2nd gnathopod as in figure. Basis of legs 5–7-denticulated. Profile of metasome with sharp tooth on 1st and 2nd segments. 3rd epimeral plate with infero-posterior corner; posterior border distinctly toothed. Cleft on telson slightly exceeding 1/3 of length.

### Remarks

All 4 juvenile specimens were characterized by the tooth on the 1st and 2nd metasomal segments, the strongly denticulated epimeral plate 3, the denticulated basis of legs 5–7, and the peduncle of the antennulae, the first two articles of which were equal in length.

### Distribution

Bellan-Santini (1972) reports the following distribution for the species: South America, and Antarctic and sub-Antarctic waters, where it occurs from 0–547 m. In the Straits of Magellan, it was sampled in very low numbers at St. 21 at the mouth of the Primera Angostura, where it occurred from the surface to the 40 m depth.

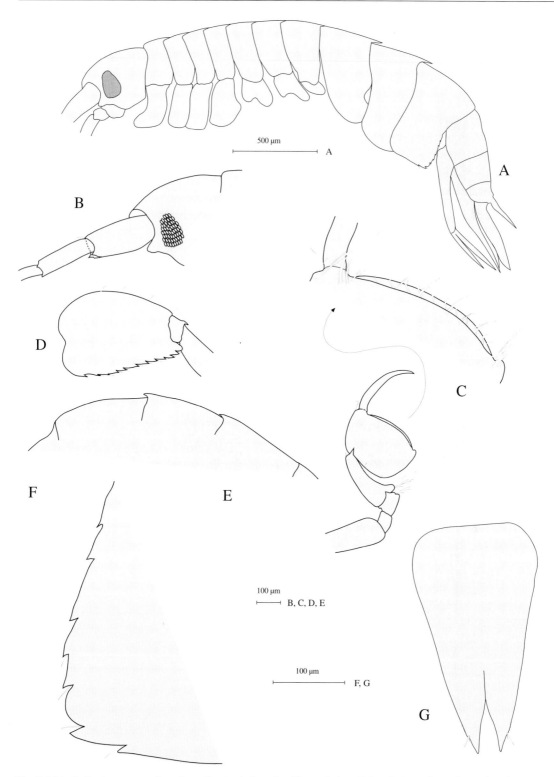

**Fig. 3.4.1A–G.** *Eusirus antarcticus.* Juvenile: **A** whole animal lateral view; **B** head; **C** gnathopod 2 with palmar margin further enlarged; **D** 2nd article of pereopod 7; **E** dorsal profile of pleon segments 1–3; **F** 3rd epimeral plate; **G** telson

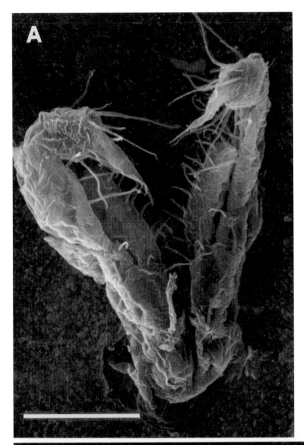

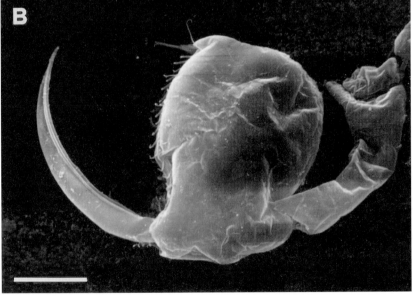

**Fig. 3.4.2A, B.** *Eusirus antarcticus.* Juvenile: **A** maxilliped; **B** distal portion of gnathopod 2. *Bars* **A,B** 100 μm

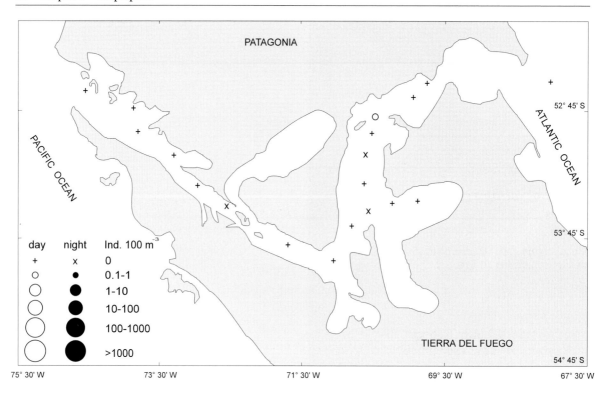

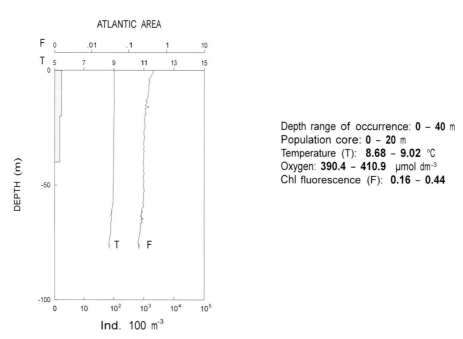

Depth range of occurrence: **0 – 40** m
Population core: **0 – 20** m
Temperature (T): **8.68 – 9.02** °C
Oxygen: **390.4 – 410.9** μmol dm⁻³
Chl fluorescence (F): **0.16 – 0.44**

**Fig. 3.4.3.** *Eusirus antarcticus.* Distribution in the Straits of Magellan in February–March 1991

## 3.5 *Rhachotropis schellenbergi* Andres, 1982

*Rhachotropis* sp. Schellenberg, 1931: p. 173.
*Rhachotropis schellenbergi* Andres, 1982: p. 174, figs. 12–15a.

Body length –
    males: 9.6–10.0 mm (8 specimens)
    females: 7–9 mm (6 specimens)

**Male.** Mesosome smooth; 1st to 3rd segments of metasome tricarinate. Median keel of 3rd metasomal segment rounded posteriorly. 1st epimeral plate lacking teeth. 3rd epimeral plate rounded and serrated. 1st coxal plate protruding forward. 1st segment of flagellum on antennulae bearing numerous aesthetascs. 2nd antennae with last two articles of peduncle bearing calceoli. 1st and 2nd pereopods similar to gnathopods. 3rd and 4th pereopods slender and equal in length; 5th and 6th pereopods very thin. Length and width of the basis increasing from 5th to 7th pereopods. Outer rami of 1st and 2nd uropods shorter than inner rami. Rami of 3rd uropod almost of the same length. Margins of uropodal rami with small spines. Telson long and cleft up to almost 1/2 its length; both lobes notched terminally and bearing one seta each. Plumose seta inserted proximally on both sides.

**Female.** General shape as in male, except for the shorter antennae and the scarcity of 1st antennae aesthetascs.

### Remarks

According to Andres (1982), the rounded 1st epimeral plate and the absence of teeth on the 1st urosomal segment allow the easy identification of the species from the congeneric species *R. antarctica*.

### Distribution

Andres (1982) reports the locus typicus of the species as the Antarctic Ocean (61°29'S, 56° 20'W) at a depth of 325 m. In the Straits of Magellan, it was present only in the area of Paso Ancho with a vertical depth range from 50–200 m (St. 19).

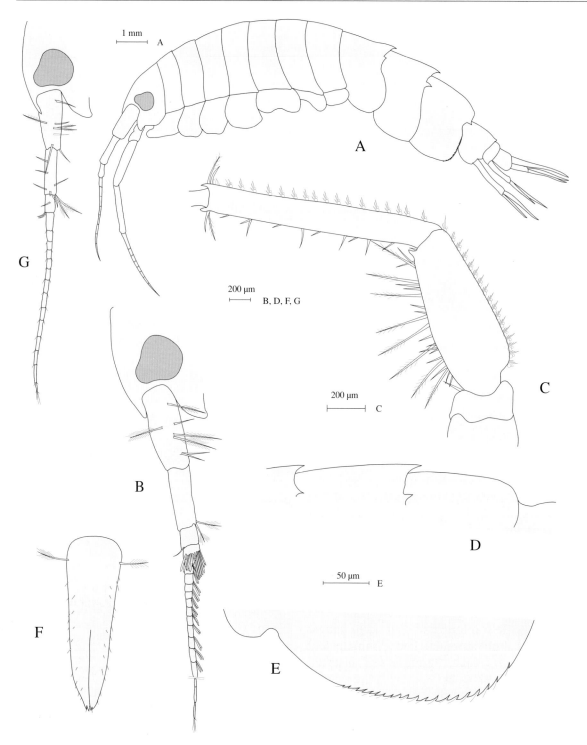

**Fig. 3.5.1A–G.** *Rhachotropis schellenbergi.* Male: **A** whole animal lateral view; **B** head with proximal and distal portion of antennae 1; **C** antennae 2; **D** latero-dorsal profile of pleon; **E** 3rd epimeral plate; **F** telson. Female: **G** head with antennae 1

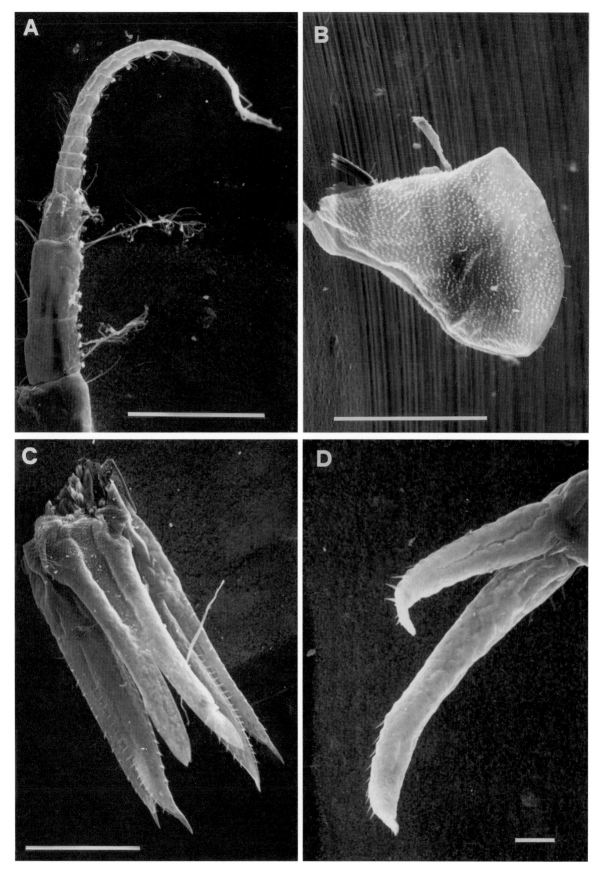

**Fig. 3.5.2A–D.** *Rhachotropis schellenbergi*. Female: **A** antenna 1; **B** 1st epimeral plate; **C** dorsal view of urosome with uropods and telson; **D** uropod 2. *Bars* **A,B,C** 500 μm; **D** 100 μm

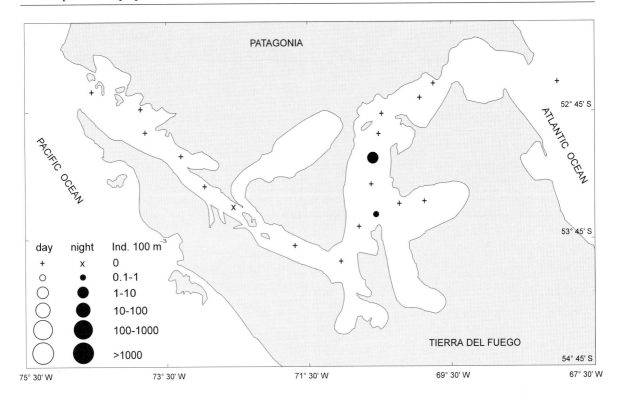

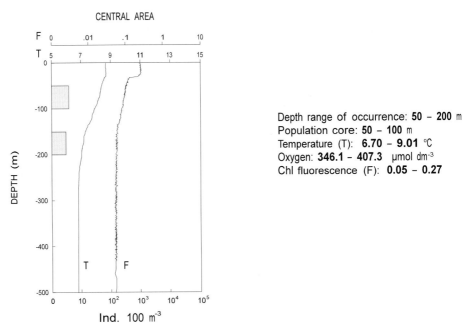

**Fig. 3.5.3.** *Rhachotropis schellenbergi.* Distribution in the Straits of Magellan in February–March 1991

## 3.6 *Jassa alonsoae* Conlan, 1990

*Jassa alonsoae* Conlan, 1990: p. 2045, fig. 13.
*Jassa alonsoae* De Broyer and Jazdzewski, 1993: p. 59.

Body length –
   female: 5.9 mm (1 specimen)

**Female.** Head bearing small, protruding, lateral lobes with dorsal angle more acute. A1 shorter than A2, with very small accessory flagellum. Gnathopod 1: basis, anterolateral margin with 1 row of short setae along its length, posterior margin with setae only at junction with ischium; carpus with setae at the anterodistal junction with propodus. 2nd gnathopod larger than 1st; basis with seate on anterolateral margin; palmar border sinuous and bearing 2 small teeth near the dactyl insertion and 3 proximal, short, stout spines. 1st and 2nd uropods with outer ramus shorter than the inner. 3rd uropod with hook-like outer ramus and 2 small teeth on upper distal margin. Telson very small, triangular in shape, with 1 small seta on either side of sharp apex.

### Remarks

The amphipod genus *Jassa*, since its description in 1814, has caused considerable difficulty in identification between carcinologists. On the basis of cladistic and phenetic analysis of 141 characters, Conlan (1989) concluded that the genus *Jassa* should comprise 19 species. *Jassa alonsoae*, according to Conlan (1990), is characterized by the combination of a setose basis of gnathopod 2, a sinuous palm, and the presence of a setal cluster at the anterodistal junction of the carpus and propodus on gnathopod 1.

### Distribution

Conlan (1990) reports the following sampling locations: Puerto Deseado (Southern Argentina), Tierra del Fuego (Southern Chile), Falkland Islands, South Georgia (Tristan da Cunha), Kerguelen Island, Hauckland Islands, Macquarie Island, The Snares Islands. Depth range, low intertidal to 150 m. In the present study, it was very rare and was sampled only at St. 22 from 10–20 m.

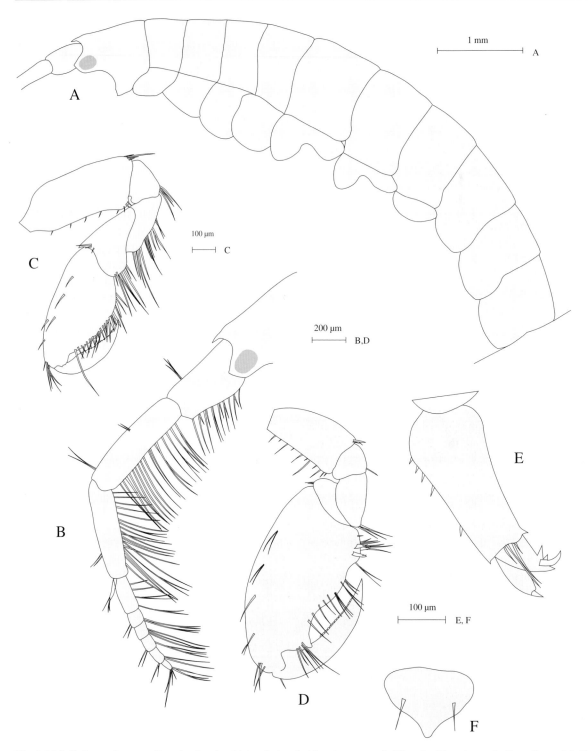

**Fig. 3.6.1A–F.** *Jassa alonsoae.* Female: **A** animal lateral view (without urosome); **B** head with antenna 1; **C** gnathopod 1; **D** gnathopod 2; **E** uropod 3; **F** telson

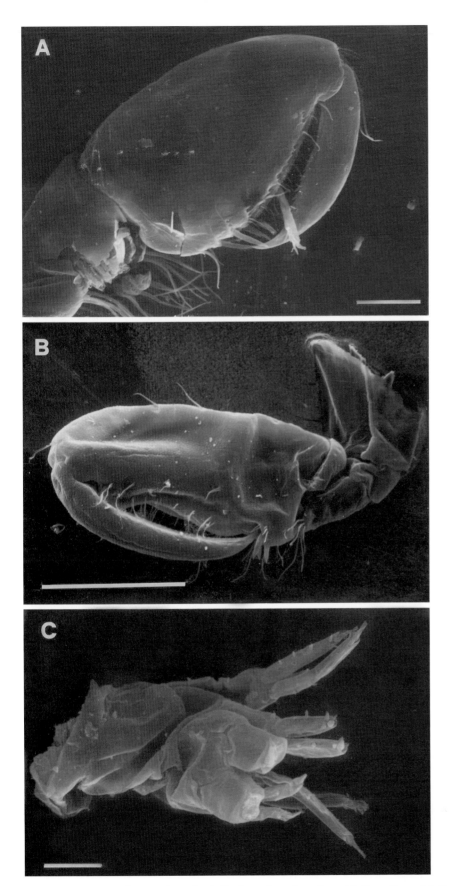

**Fig. 3.6.2A–C.** *Jassa alonsoae*. Female: A distal portion of gnathopod 1; B gnathopod 2; C urosome. *Bars* A,C 100 μm; B 500 μm

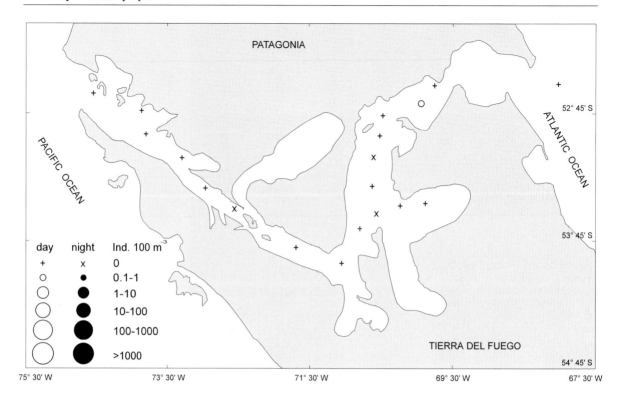

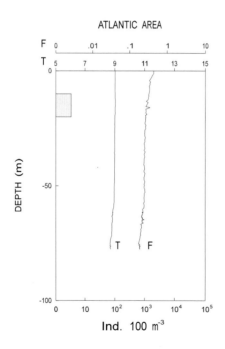

Depth range of occurrence: **10 – 20** m
Population core: **10 – 20** m
Temperature (T): **8.68 – 9.02** °C
Oxygen: **390.4 – 410.9** µmol dm⁻³
Chl fluorescence (F): **0.16 – 0.44**

**Fig. 3.6.3.** *Jassa alonsoae*. Distribution in the Straits of Magellan in February–March 1991

### 3.7 *Scina borealis* Sars, 1883

*Clydonia borealis* Sars, 1883: p. 76, tab. 3 fig. 1a–b.
*Scina borealis* Sars, 1895: p. 20, pl. 8–Chevreux, 1919: p. 16–Wagler, 1926: p. 337, figs. 9–11–Dick, 1970: p. 29 (key)– Vinogradov et al., 1982: p. 146, fig.65.

Body length –
    females: 2.6–6.0 mm (4 specimens)
    males: 4.9–5.3 (3 specimens)

**Female.** Head deeper than long. A2 small and filiform. 1st maxilla with outer lobe that is strongly serrated at apex and hairy at the base; inner lobe smooth and bearing small apical tooth. Maxilliped with well-developed, lanceolated outer lobes; inner lobes entirely fused. 1st pereopod with carpus longer than metacarpus. 2nd pereopod with carpus and metacarpus almost of the same length. 5th pereopod with basis bearing spines on either margin and terminating with sharp process that overhangs ischial segment; metacarpus smaller than carpus. 6th pereopod shorter than 5th. 7th pereopod small and weak. Outer rami of 1st and 2nd uropods spine-like; outer rami of last uropod as long as peduncle. 1st uropod ornated with fine teeth on outer margin; inner margin bearing strong spines along entire length. 2nd uropod with smooth outer margin and denticulated inner margin similar to outer margin of inner ramus and inner margin of outer ramus of 3rd uropod. Telson small and acutely triangular.

**Male.** General body shape as female. A2 as long as body, with elongated last peduncle article.

**Remarks**

According to Wagler (1926), the main characteristics of the species include the following: maxilliped with well-developed, sharp external lobes; 5th pereopod with merus equal to carpus; propodus reduced compared to carpus; basipod bearing stout spines on either side; 1st uropod with long spines on inner margin; 3rd uropod with outer ramus almost equal to peduncle.

**Distribution**

Chevreux (1919) reports that the Prince of Monaco collected the species in the Atlantic Ocean and Mediterranean Sea. The species has also been sampled in the Indian Ocean as well as in Antarctic waters. It is considered a pan-oceanic species with an ample vertical distribution range (Thurston 1976). Bogorov (1958) considers *S. borealis* as a characteristic amphipod in the 200–500-m-depth layers of the NW boreal Pacific Ocean. Vinogradov (1970) reports a depth range for the species from mesopelagic (200–500 m) to bathypelagic (2000–3000 m) waters. In the Straits of Magellan, it was found only at 2 very deep Pacific stations. *S. borealis* was distributed at all depths from the surface to 700 m.

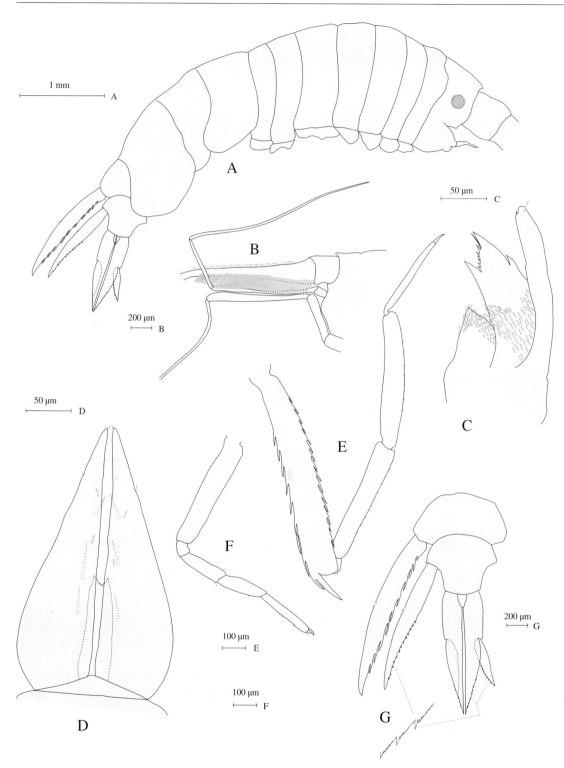

**Fig 3.7.1A–G.** *Scina borealis.* Female: **A** whole animal lateral view; **B** head of male with antennae 1and 2; **C** maxilla 1; **D** maxilliped; **E** pereopod 5; **F** pereopod 7; **G** dorsal view of urosome with uropods and telson

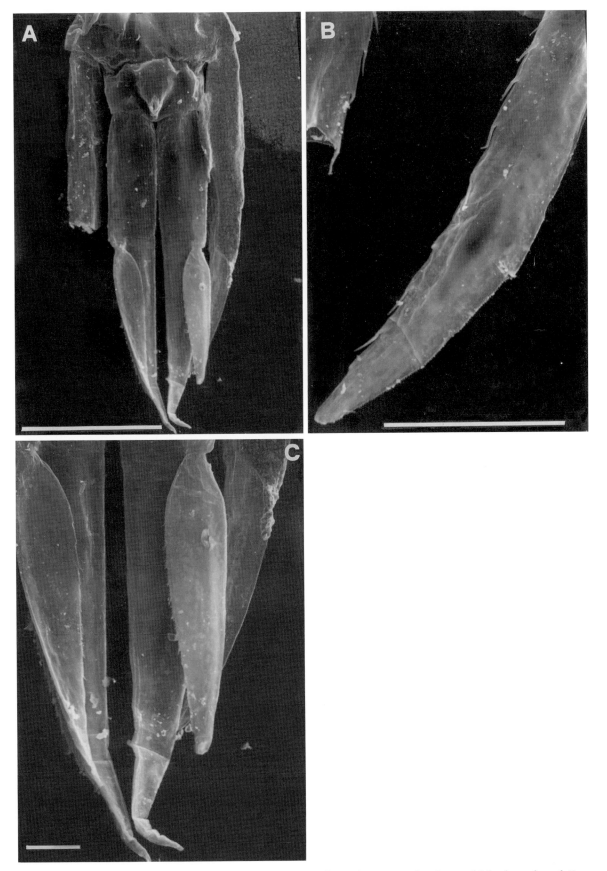

**Fig. 3.7.2A–C.** *Scina borealis.* Male: **A** urosome (part) with uropods 2 and 3; **B** uropod 1; **C** uropod 3 further enlarged. *Bars* **A,B** 500 μm; **C** 100 μm

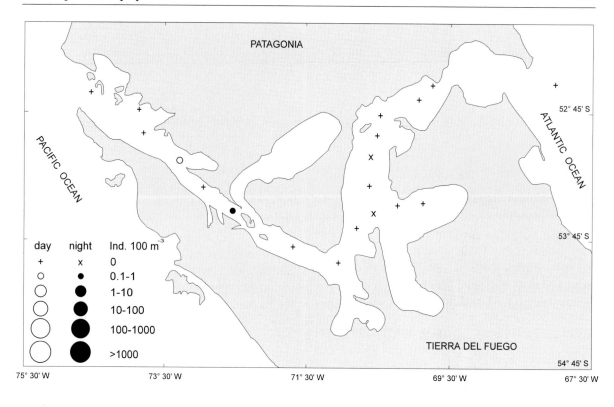

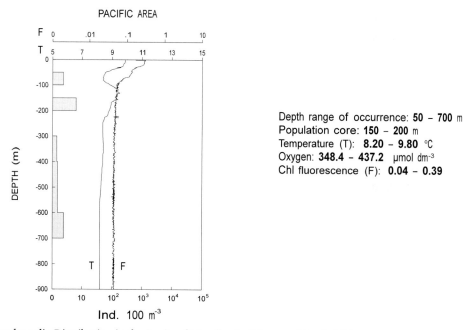

**Fig. 3.7.3.** *Scina borealis.* Distribution in the Straits of Magellan in February–March 1991

## 3.8 *Cyllopus magellanicus* Dana, 1853

*Cyllopus magellanicus* Dana, 1853: p. 990, pl. 68 fig. 1 – Barnard, 1930: p. 408 – Barnard, 1932: p. 266 – Hurley, 1955: p. 129, figs. 23–50 – Weigmann-Haass, 1983: p. 2 (key), figs. 1–3.
*Vibilia macropis* Bovallius, 1887: p. 51, pl. 8 figs. 1–8.
*Cyllopus levis* Bovallius, 1889: p. 8, pl. 1 figs. 36–41.
*Cyllopus armatus* Bovallius, 1889: p. 10, pl. 1 figs. 1–35.

Body length –
   female: 6.6 mm (1 specimen)
   male: 7.2 mm (1 specimen)

**Female.** Large eyes that cover almost the entire head. A1 shorter than A2. A1 with short, stout 1st article of the flagellum followed by two small articles. P1 simple; propod and dactyl with slightly denticulated posterior border. P2 with strongly sinulated border of merus; carpus, propod and dactyl with serrated margins. P7 with oval base and convex posterior margin. Rami of 3rd pair of uropods acutely pointed and with serrated margins.

**Male.** Similar in body shape to female except for 1st article of flagellum of A1, which is longer and thinner in males.

**Remaks**

The genus *Cyllopus* at present includes two species, *C. magellanicus* and *C. lucasii*. The distribution of both is confined to Antarctic waters. The presence of two additional segments at the end of the 1st antenna and the convex posterior margin of the P7 allows the easy identification of *C. magellanicus* from the congeneric species, *C. lucasii*.

**Distribution**

*C. magellanicus* is widely distributed from the Antarctic to South Georgia. Its distribution range extends from the Antarctic continent up to the subtropical convergence (Weigmann-Haass 1983). According to Barnard (1930), Bary (1959), Hurley (1960), Kane (1962), Vinogradov (1966) and Semenova (1976), *C. magellanicus* is present also in the Antarctic sector of the Pacific and Indian Oceans. The species therefore has a circumpolar distribution, being limited in its diffusion by the intensity of the Western winds. In the Straits of Magellan, it was sampled exclusively from 100–200 m at the mouth of the Pacific entrance.

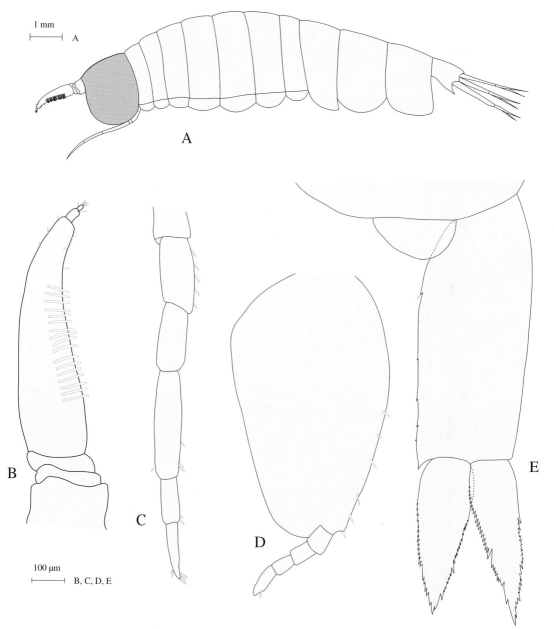

**Fig. 3.8.1A–E.** *Cyllopus magellanicus.* Female: **A** whole animal lateral view (Modified after Ramirez and Viñas 1985); **B** antennae 1; **C** antennae 2; **D** pereopod 7; **E** telson and uropod 3

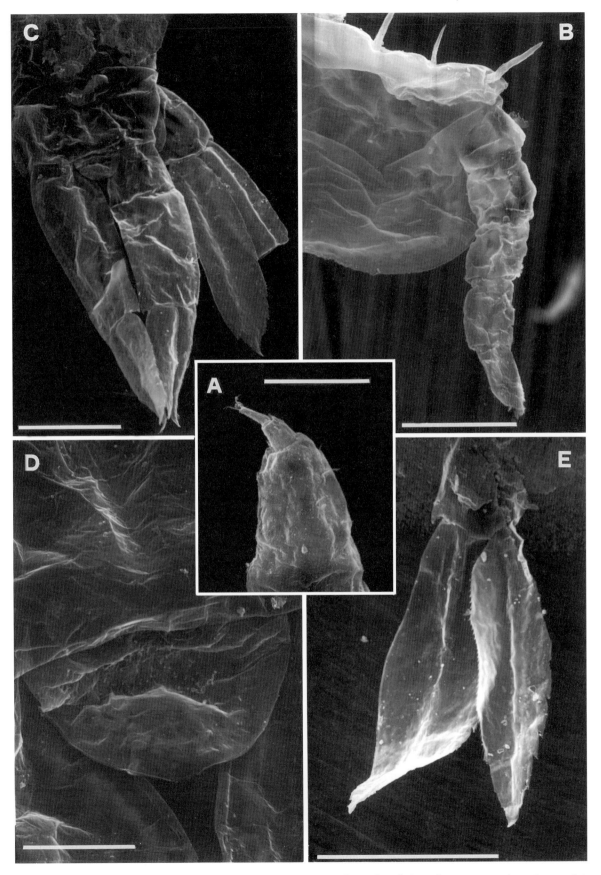

**Fig. 3.8.2A–E.** *Cyllopus magellanicus*. Male: **A** antennae 1; **B** pereopod 7; **C** dorsal view of urosome; **D** telson; **E** uropod 2.
*Bars* **A,B,D** 100 µm; **C,E** 500 µm

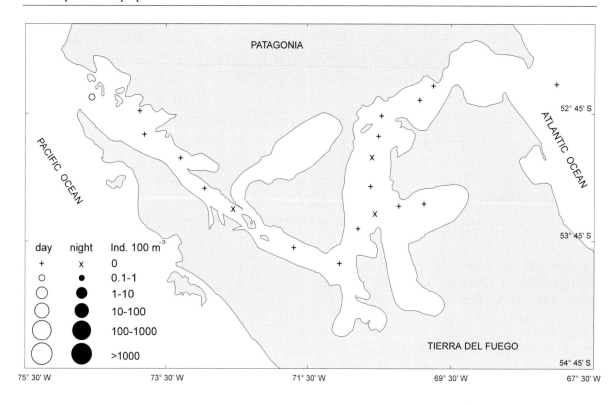

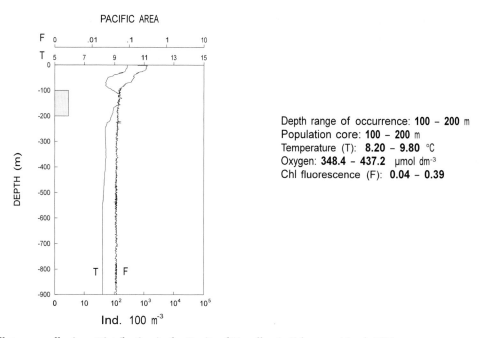

**Fig. 3.8.3.** *Cyllopus magellanicus.* Distribution in the Straits of Magellan in February–March 1991

## 3.9   *Hyperiella dilatata* Stebbing, 1888

*Hyperiella dilatata* Stebbing, 1888: p. 1403, pl. 171 – Bovallius, 1889: p. 247 – Barnard, 1932: p. 274, fig. 161 – Hurley, 1969: p. 33 – Bowman, 1973: p. 27, figs. 20a–m, 21a–f – Weigmann-Haass, 1989: p. 184, figs. 34–43, p. 181 (key).

Body length –
   immature male: 4.1 mm (1 specimen)

**Male.** Both A1 and A2 with multisegmented flagellum. 1st pereopod with subchela; 2nd pereopod with chela. 5th pereopod much longer than 6th and 7th. Epimeral plates of pleon with pointed postero-lateral corners. 3rd uropod with exopod 2/3 as wide as endopod.

### Remarks

Bowman (1973) maintains that the diagnostic key proposed by Bovallius (1889) to distinguish *H. dilatata* from *H. antarctica* is based on characters that are poorly reliable. More recently, Weigmann-Haass (1989) provided a key to identify 3 Antarctic species, *H. antarctica* Bovallius 1887, *H. dilatata* Stebbing 1888 and *H. macronix* (Walker, 1906). This key allows the easy identification of *H. dilatata* from the two other congeneric species.

### Distribution

Bowman (1973) reports the Indian sector of the Antarctic (63°30'S, 88°57'E) as the type locality of *H. dilatata*, but the species has also been sampled in the Ross Sea. More recently, Weigmann-Haass (1989) considers the species as circumpolar in the Antarctic Ocean. In the Straits of Magellan, it was found only at St. 5, at the mouth of the Pacific entrance, where it was sampled from 50–100 m.

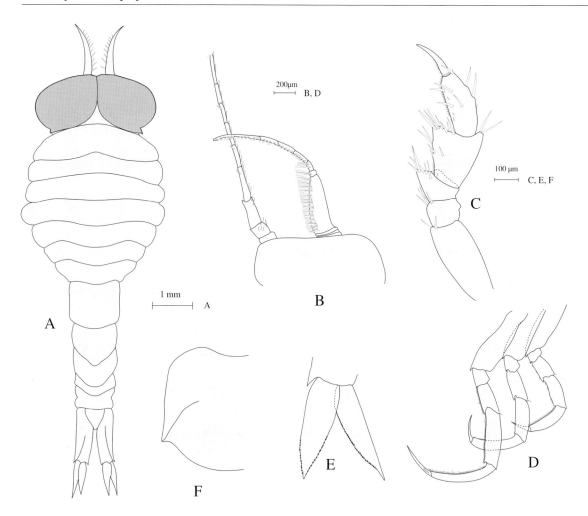

**Fig. 3.9.1A–F.** *Hyperiella dilatata.* Male: **A** whole animal dorsal view (Modified after Stebbing 1888); **B** head with antennae 1 and portion of antennae 2; **C** pereopod 1 (= gnathopod 1); **D** pereopods 5, 6 and 7; **E** rami of uropod 3; **F** 2nd epimeral plate

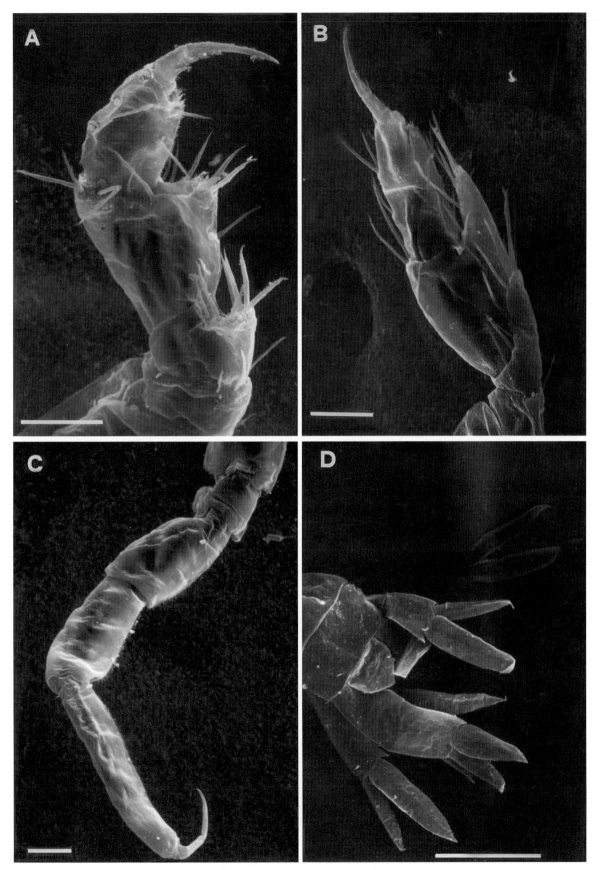

**Fig. 3.9.2A–D.** *Hyperiella dilatata*. Male: **A** distal portion of pereopod 1; **B** distal portion of pereopod 2; **C** pereopod 5; **D** dorsal view of urosome. *Bars* **A,B,C** 100 μm; **D** 500 μm

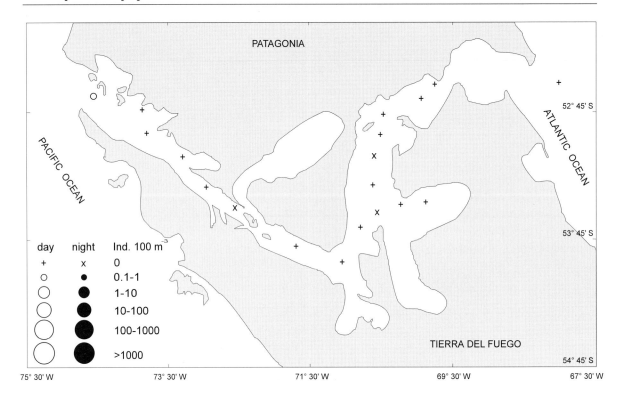

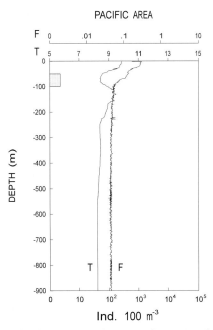

Depth range of occurrence: **50 – 100** m
Population core: **50 – 100** m
Temperature (T): **8.20 – 9.80** °C
Oxygen: **348.4 – 437.2** μmol dm⁻³
Chl fluorescence (F): **0.04 – 0.39**

**Fig. 3.9.3.** *Hyperiella dilatata.* Distribution in the Straits of Magellan in February–March 1991

### 3.10 *Hyperoche mediterranea* Senna, 1908

*Hyperoche mediterranea* Senna, 1908: p. 159 – Barnard, 1930: p. 415 (key) – Hurley, 1955: p. 147, figs. 115–132.

Body length –
   female: 4.7 mm (1 specimen)

**Female.** Left mandible with incisor consisting in a narrow scalpel-shaped process with serrated margins. Molar process with a wide, triangular blade projecting inwards. Between the two is a small, finely setous, subtriangular process in correspondence to the row of spines. Palp long, three-segmented; 3rd segment covered with fine setules. Inner lobe of 1st maxilla encircled by setae; oval lamellar outer plate, with inner margin bearing few setules and minute teeth on distal extremity. Inner lobe of 2nd maxilla pointed and covered with fine setules; outer lobe similar to a thin, setous palp. Maxilliped with thin median, finely setous process extending almost to half outer lanceolated plate. This plate bears slightly setous inner margin and pointed apices. 1st and 2nd gnathopods with posterior margins of merus and carpus prolonged into sharp processes. Basis, ischium, merus, carpus and propodus bearing fine spinulation along posterior margins and nearby areas. 3rd pereopod with carpus ending in pointed postero-distal angle. Epimeral plates with largely rounded postero-distal corner.

**Remarks**

The species can be easily distinguished from the congeneric species *H. medusarum* by the lanceolated form of the outer plates of the maxilliped that terminate in acute apices and the rounded edges of the epimeral plates. In *H. medusarum*, the outer plates of the maxilliped are rounded at the apex and the epimeral plates bear postero-distal corners terminating with a small point.

**Distribution**

Dick (1970) considers the species as relatively rare in temperate regions and reports that it is distributed in the SE temperate Atlantic and Pacific Oceans, as well as the Mediterranean Sea. In the Straits of Magellan, it was sampled only in surface waters at St. 5 situated at the mouth of the Pacific entrance.

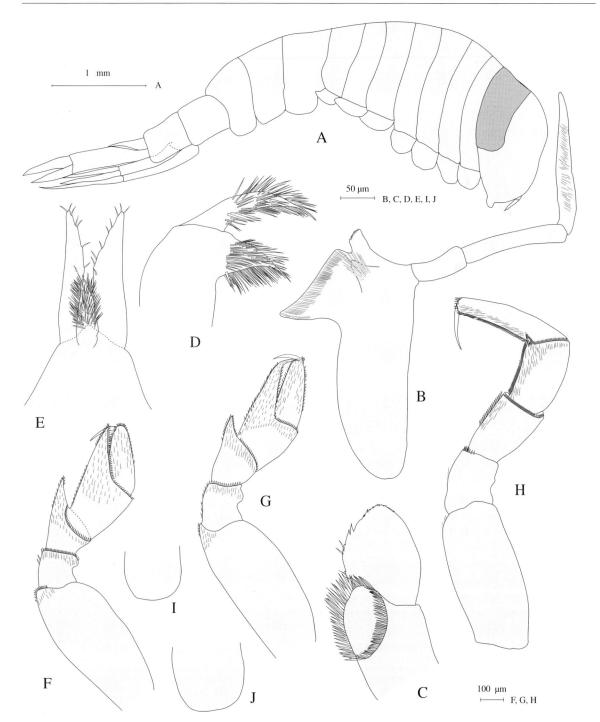

**Fig. 3.10.1A–J.** *Hyperoche mediterranea.* Female: **A** whole animal lateral view; **B** left mandible; **C** maxilla 1; **D** maxilla 2; **E** maxilliped; **F** gnathopod 1; **G** gnathopod 2; **H** pereopod 3; **I** epimeral plate 1; **J** epimeral plate 3

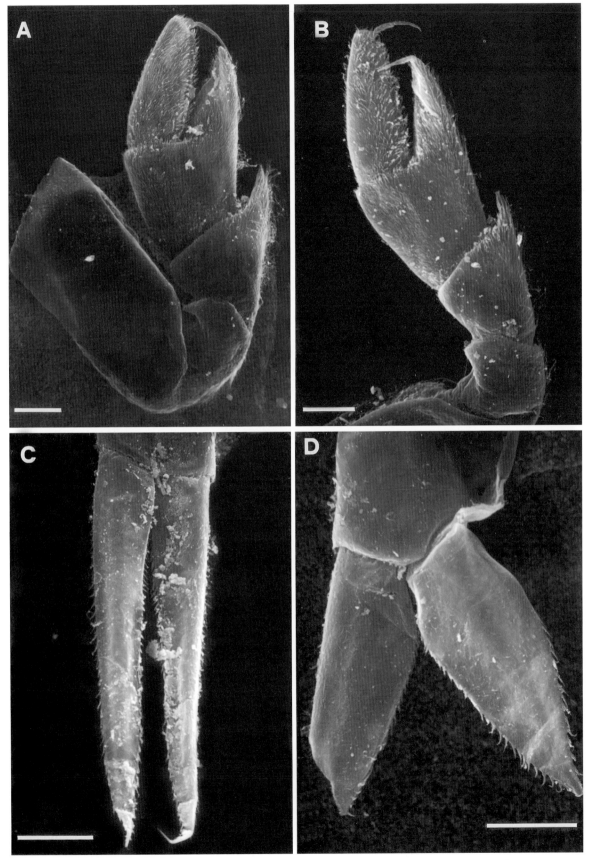

**Fig. 3.10.2A–D.** *Hyperoche mediterranea.* Female: **A** gnathopod 1; **B** gnathopod 2; **C** uropod 1; **D** uropod 3. *Bars* **A,B,C,D** 100 μm

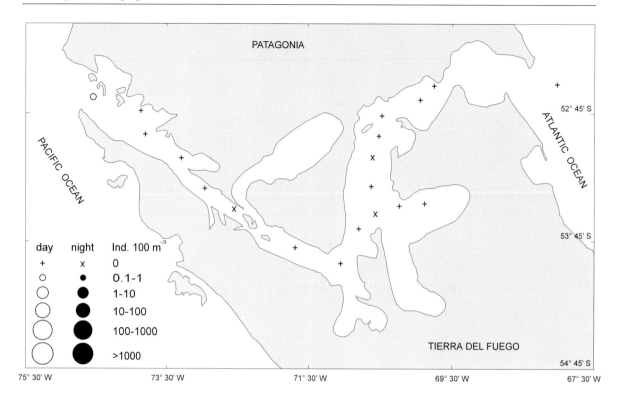

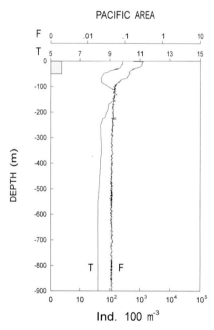

Depth range of occurrence: **0 – 50** m
Population core: **0 – 50** m
Temperature (T):  **8.20 – 9.80** °C
Oxygen: **348.4 – 437.2**  μmol dm⁻³
Chl fluorescence (F): **0.04 – 0.39**

**Fig. 3.10.3.** *Hyperoche mediterranea.* Distribution in the Straits of Magellan in February–March 1991

## 3.11 *Hyperoche medusarum* Kröyer, 1838

*Metoecus medusarium* Kröyer, 1838: p. 288.
*Hyperoche kroeyeri* Bovallius, 1889: pp. 87–92.
*Hyperoche medusarum* Barnard, 1930: p. 415 (key) – Hurley, 1955: p. 144, figs. 96–114.

Body length –
    females: 4.5–8.5 mm (4 specimens)
    males: 3.4–6.2 mm (2 specimens)

**Female.** A1 and A2 bearing 1-articulate flagellum. Incisor of left mandible consisting in two processes bearing small distal teeth; molar process triangular and turned inwards. Incisor of right mandible with one process bearing small distal teeth. Inner lobe of 1st maxilla oval and encircled by setules; outer plate lamellar with inner margin ornated by short setules and distally bearing minute teeth. Outer plates of maxilliped oval with convex outer margins and rounded apex; median lobe with oval plate, surface of which bears fine setules. P1 and P2 with distal chelae. Epimeral plates with slightly convex posterior margin; postero-distal corners acute. A fine and distally serrated suture forms an arc from the postero-distal corner of each plate to the antero-proximal.

**Male.** General form as in female, but more slender. A1 and A2 much longer than in female. P1 and P2 with narrow carpus.

### Remarks

The species can be easily distinguished from the congeneric species *H. mediterranea* due to the rounded apex of the outer lobes of the maxilliped and the epimeral plates with postero-distal corners protruding into a small point.

### Distribution

Hurley (1955) reports the following sampling locations: N. Atlantic and adjacent seas, north of Alaska, South Georgia and New Zealand. Bowman et al. (1963) gives its presence also in the NW Atlantic, Chesapeake Bay. More recently, Raymont (1983) describes the species as boreal-arctic and as present in cold waters of both oceans. In the Straits of Magellan, the species was not sampled in the Atlantic sector, whereas it was common in the Central area, from Cape Froward to Paso Ancho. Its vertical distribution range in Central waters was from 150 m to the bottom.

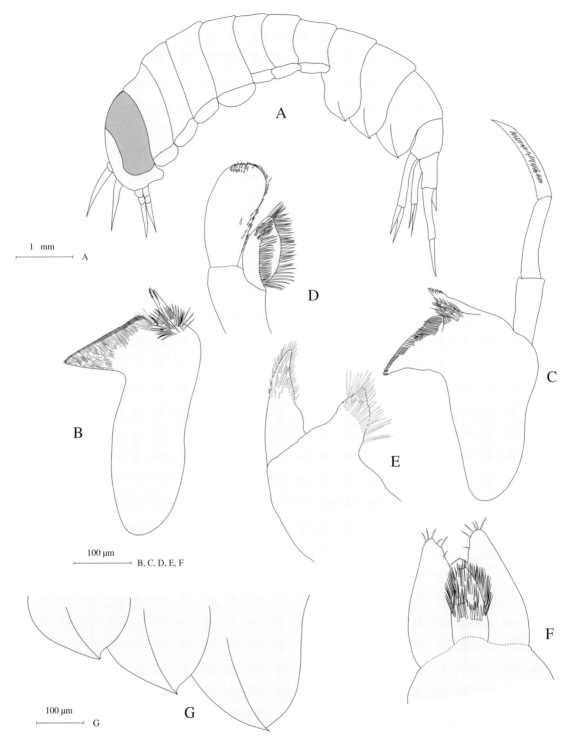

1 mm
A

100 µm
B, C, D, E, F

100 µm
G

**Fig. 3.11.1A–G.** *Hyperoche medusarum*. Female: **A** whole animal lateral view (Modified after Hurley 1955); **B** left mandible; **C** right mandible; **D** maxilla 1; **E** maxilla 2; **F** maxilliped; **G** epimeral plates 1–3

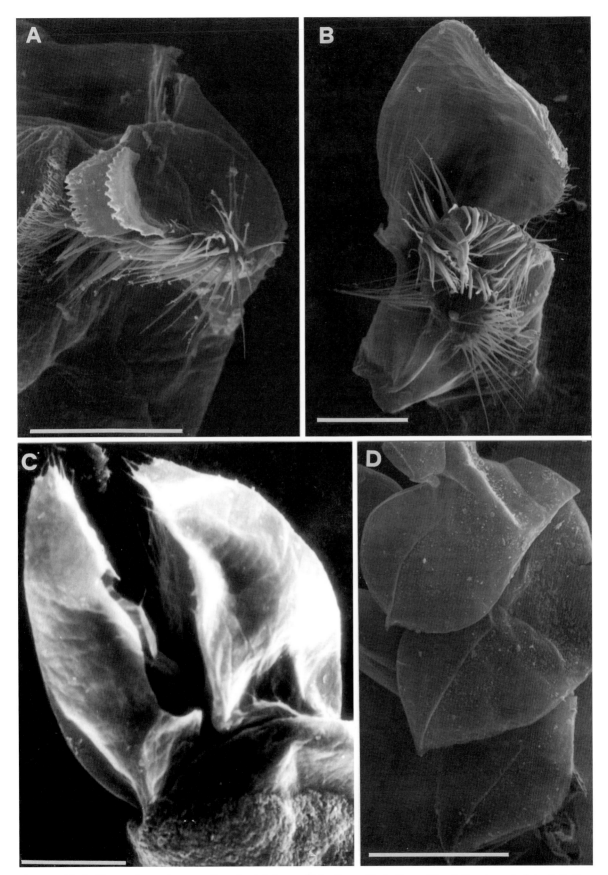

**Fig. 3.11.2A–D.** *Hyperoche medusarum.* Female: **A** left mandible; **B** maxilla 1; **C** maxilliped; **D** epimeral plates 1–3. *Bars* A,B,C 100 μm; D 500 μm

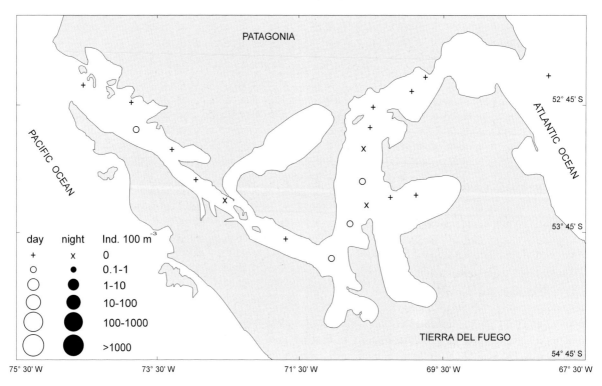

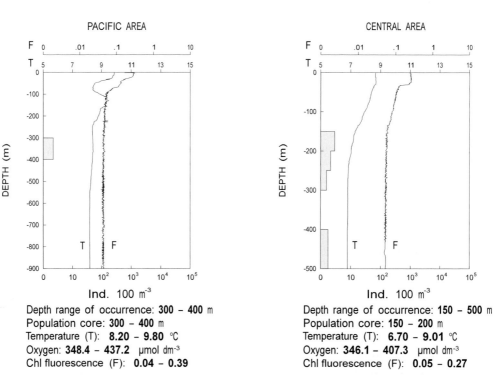

Depth range of occurrence: **300 – 400** m
Population core: **300 – 400** m
Temperature (T): **8.20 – 9.80** °C
Oxygen: **348.4 – 437.2** µmol dm⁻³
Chl fluorescence (F): **0.04 – 0.39**

Depth range of occurrence: **150 – 500** m
Population core: **150 – 200** m
Temperature (T): **6.70 – 9.01** °C
Oxygen: **346.1 – 407.3** µmol dm⁻³
Chl fluorescence (F): **0.05 – 0.27**

**Fig. 3.11.3.** *Hyperoche medusarum.* Distribution in the Straits of Magellan in February–March 1991

## 3.12 *Themisto gaudichaudii* Guérin, 1828

*Themisto gaudichaudii* Guérin, 1828: p. 774
*Euthemisto gaudichaudii* Bovallius, 1889: p. 299,
figs. 1–5, pl. 13 figs. 44–46.
*Parathemisto (Euthemisto) gaudichaudii* Hurley,
1955: p. 161, figs. 159–174.
*Parathemisto gaudichaudii* Sheader and Evans,
1974: p. 915 (key) – Sheader, 1975: p. 887 –
Schneppenheim and Weigmann-Haass, 1986: p.
219, figs. 1, 1a, p. 222 (key).

Body length –
    females: 6.4–13.0 mm (30 specimens)
    males: 7.0–13.0 mm (30 specimens)

**Female.** Body keel-like dorsally with strong dorsal spines about ≥ 10 mm in length. Article 4 of A1 curved. Flagellum of A2 with one article. Anterior extremity of P2 merus rounded. P3 and P4 with dilated carpus that together with propod, form a prehensile organ. P5 much longer than P6 and P7 in the bispinosa form and only slightly longer than the other two in the compressa form.

**Male.** Peduncle of A1 3-segmented; multisegmented flagellum with 1st segment bearing two rows of aesthetaecs. A2 longer than A1 with multisegmented flagellum.

### Remarks

This was the only species sampled at all stations and was also the most numerically abundant amphipod in the Straits. Males were rarer than females in both the bispinosa and compressa forms. Barnard (1932) also reported a small percentage (2.3 %) of males in the population. The percentage of males sampled in the present study was 0.70 % in the bispinosa form and 12 % in the compressa form. In accordance with Barnard (1932) and Schneppenheim and Weigmann-Haass (1986), we also found a larger number of bispinosa (77.7 %) than compressa (22.3 %) forms. Schneppenheim and Weigmann-Haass (1986) report that: "A2 of females and males have a multisegmented flagellum". This is true only for males, since females have a single, very long article, as also shown by the same authors in fig. 1. The male A1 consists in a 3-articulate peduncle and a multiarticulate flagellum, contrary to what is reported by Schneppenheim and Weigmann-Haass (1986) and Ramirez and Viñas (1985).

### Distribution

*T. gaudichaudii* was considered as having a bipolar distribution (Bowman 1960; Kane 1966; Vinogradov et al. 1982). More recently, Schneppenheim and Weigmann-Haass (1986), on the basis of morphological and electrophoretic studies, have concluded that the species is confined to Antarctic waters and has a circumpolar distribution. In the Straits of Magellan, it was frequently sampled at all stations. It is without doubt the most abundant amphipod in the Straits, from the Pacific to Atlantic entrances, with a wide distribution range from the surface to maximum depths.

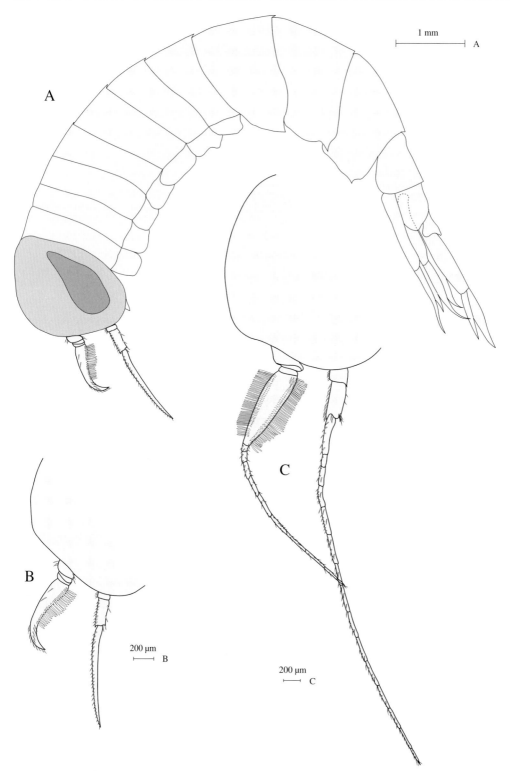

**Fig. 3.12.1A–C.** *Themisto gaudichaudii.* Female: **A** whole anima, lateral view; **B** head with antennae 1 and 2. Male: **C** head with antennae 1 and 2

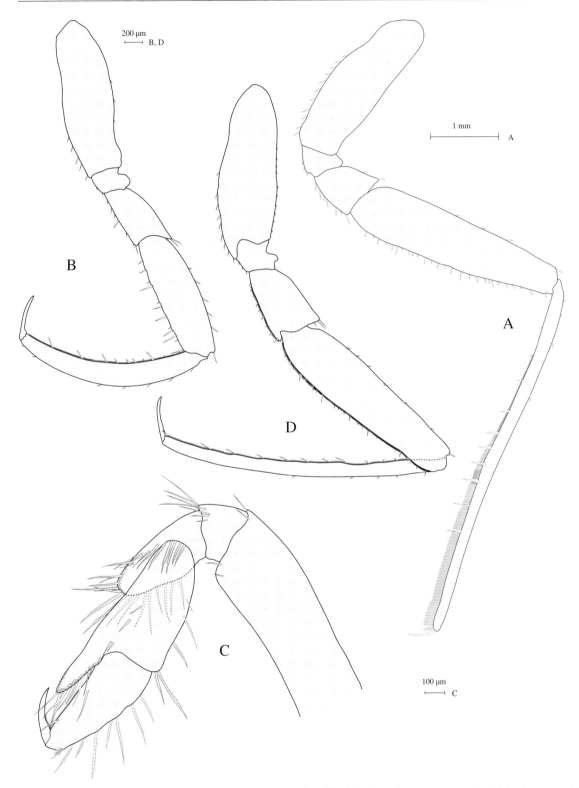

**Fig. 3.12.2A–D.** *Themisto gaudichaudii.* Female: **A** pereopod 5 of the bispinosa form; **B** pereopod 6. Male: **C** pereopod 2; **D** pereopod 5 of the compressa form

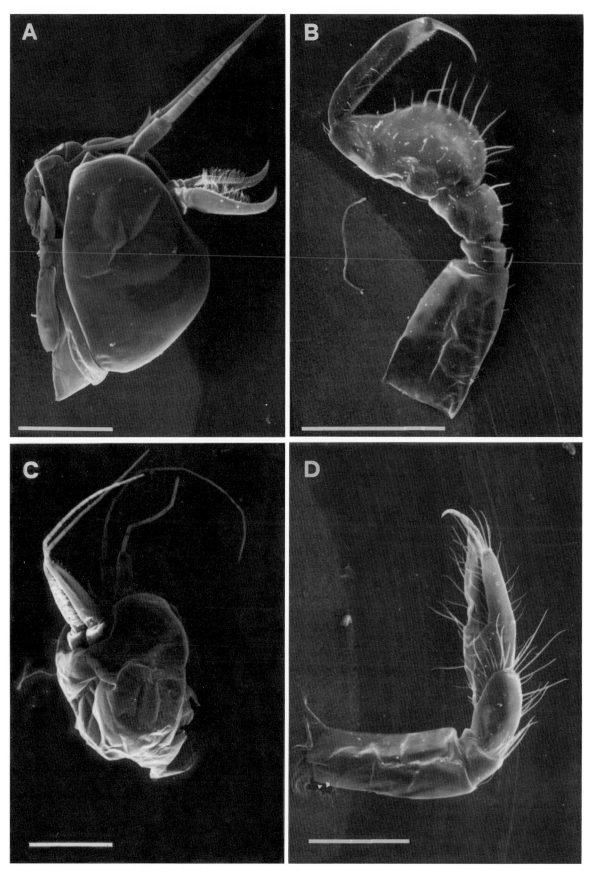

**Fig. 3.12.3A–D.** *Themisto gaudichaudii.* Female: **A** head with antennae 1 and 2; **B** pereopod 4. Male: **C** head with antennae 1 and 2; **D** pereopod 2. *Bars* **A,B,C** 1 mm; **D** 500 μm

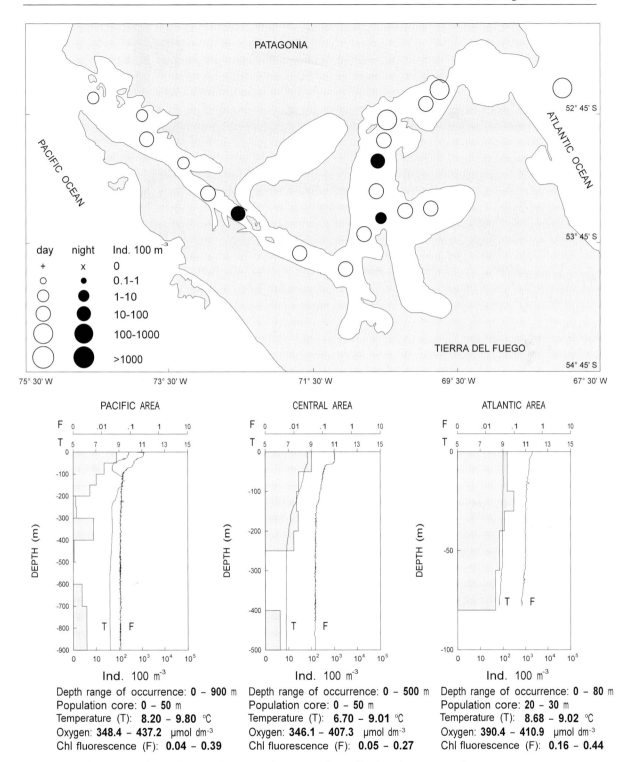

**Fig. 3.12.4.** *Themisto gaudichaudii.* Distribution in the Straits of Magellan in February–March 1991

## 3.13 *Primno macropa* Guérin-Méneville, 1836

*Primno macropa* Guérin-Méneville, 1836: p. 4, pl. 17 fig. 1a–f – Hurley, 1955: p. 172, figs. 219–235 – Bowman, 1978: p. 3, figs. 1–2, 3a–c, 4 – Bowman, 1985: p. 123, fig. 1 a–k.

Body length –
    females: 5.2–11.5 mm (6 specimens)
    male: 6.5 mm (1 specimen)

**Female.** Body compressed with evident mid-dorsal spines. Head deeper than long. A1 with 2 articles including the single pedunculated article; flagellum 1-articulate, wider proximally. A2 rudimentary. P5 with oval carpus, straight posterior margin and acute postero-distal corner; anterior margin of carpus strongly serrated with about 14 short and long teeth; proximal tooth shorter. P6 propodus armed with spine on anterior margin and minute spinules on posterior margin. P7 with posterior margin of basis slightly concave; anterior margin convex and with very rough surface; digitiform dactylum with ring of apical spines. 3rd epimeral plate with acute postero-distal corner, preceded by a feeble concavity.

**Male.** A1 with long, filiform flagellum. A2 similar to A1, but longer.

### Remarks

According to Bowman (1978), the shorter proximal tooth on the anterior margin of the carpus of the P5 and the spinous margins of the P6 propus render the identification of *P. macropa* relatively easy.

### Distribution

The species is almost cosmopolitan in its distribution (Zumelzu 1980), even though it prefers subtropical areas (Dick 1970). According to Thurston (1976), it is lacking only in the Arctic basin. In the Straits of Magellan, it was frequently sampled at the mouth of the Pacific entrance. It was not found beyond St. 11 near Carlos III Island. In all areas where it was sampled, it had a wide distribution range, from the surface to 500 m.

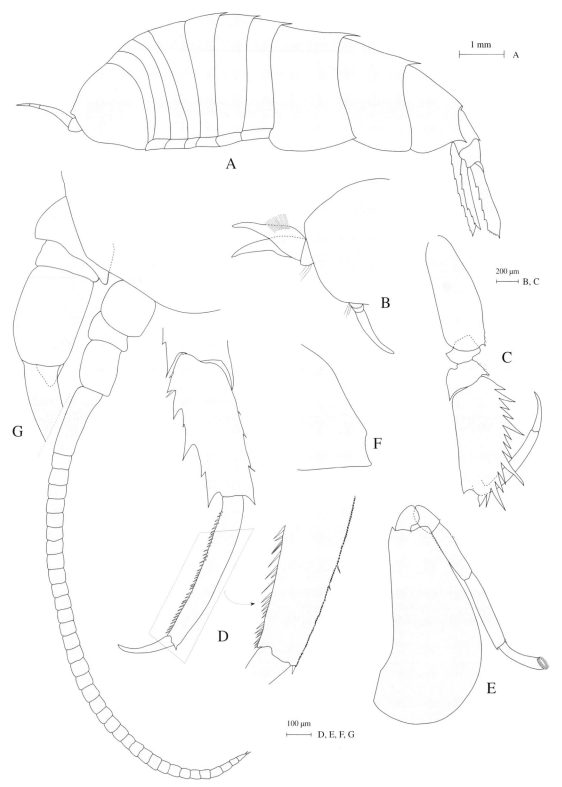

**Fig. 3.13.1A–G.** *Primno macropa.* Female: **A** whole animal lateral view (Modified after Ramirez and Viñas 1985); **B** head with antennae 1 and 2; **C** pereopod 5; **D** pereopod 6 with portion of propod further enlarged; **E** pereopod 7; **F** postero-ventral corner of 3rd epimeral plate. Male: **G** peduncular articles of antennae 1 and antennae 2

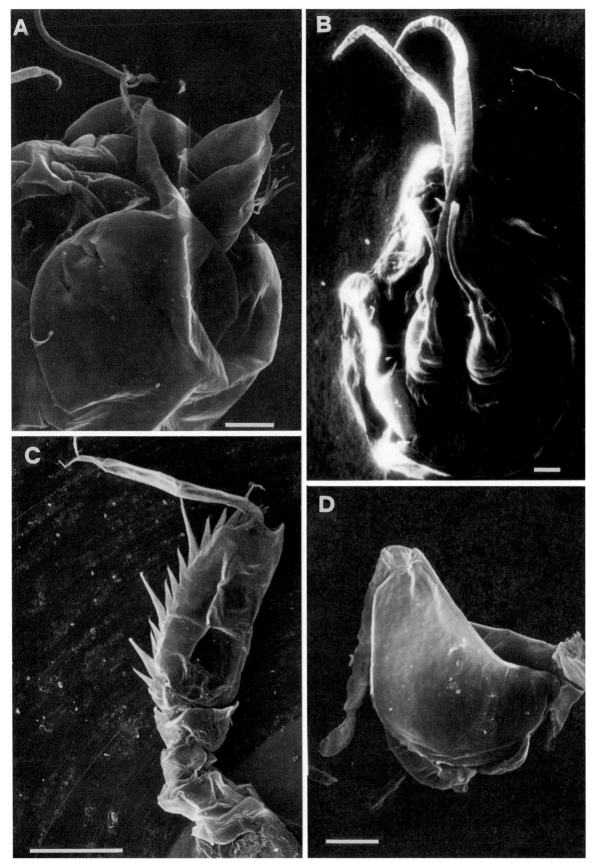

**Fig. 3.13.2A–D.** *Primno macropa.* Female: A head with antennae 1 and 2. Male: B head with antennae 1 and 2; C pereopod 5 with oval carpus strongly serrated; D pereopod 7. *Bars* **A,B,D** 100 μm; **C** 500 μm

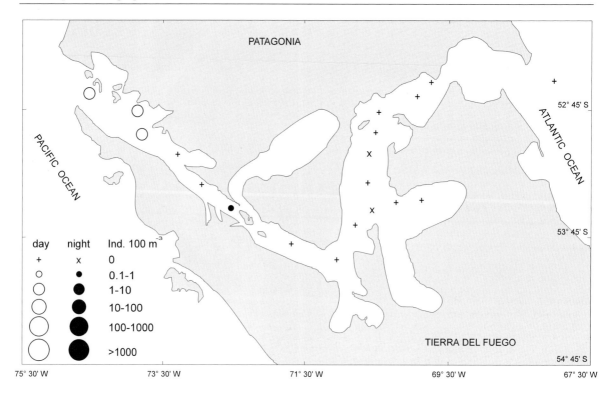

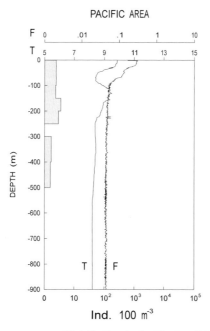

PACIFIC AREA

Depth range of occurrence:  **500 – 0** m
Population core:  **200 – 150** m
Temperature (T):  **8.20 – 9.80** °C
Oxygen:  **348.4 – 437.2** μmol dm⁻³
Chl fluorescence (F):  **0.04 – 0.39**

**Fig. 3.13.3.** *Primno macropa*. Distribution in the Straits of Magellan in February-March 1991

Andres HG (1982) Die Gammaridea (Crustacea: Amphipoda) der Deutschen Antarktis-Expeditionen 1975/76 und 1977/78, 2. Eusiridae. Mitt Hamb Zool Mus Inst 79: 159–185

Andres HG (1983) Die Gammaridea (Crustacea: Amphipoda) der Deutschen Antarktis-Expeditionen 1975/76 und 1977/78, 3. Lysianassidae. Mitt Hamb Zool Mus Inst 80: 183–220

Andres HG (1990) Amphipoda (Flohkrebse) In: Sieg J, Waegele JH (Eds) Fauna der Antarktis. Paul Parey, Berlin, pp 133–143

Barnard JL (1961) Gammaridean Amphipoda from depths of 400 to 600 meters. Galathea Rep 5: 23–128

Barnard JL (1962) South Atlantic abyssal amphipods collected by R.V. Vema Res Ser 1: 1–78

Barnard KH (1916) Contribution to the crustacean fauna of South Africa, 5. The Amphipoda. Ann S Afr Mus 15: 105–302

Barnard KH (1930) Amphipoda British Antarctic (*Terra Nova*) Exp 1910. Nat Hist Rep Zool 8: 307–454, figs 1–63

Barnard KH (1932) Amphipoda. Discovery Rep 5:1–326

Bary BM (1959) Ecology and distribution of some pelagic Hyperiidea (Crustacea-Amphipoda) from New Zealand waters. Pac Sci 13: 317–334

Bellan-Santini D (1972) Invertébrés marins des XII et XV expéditions Antarctiques Françaises en Terre Adélie. 10. Amphipodes Gammariens. Tethys Suppl 4: 157–238

Bellan-Santini D, Ledoyer M (1974) Gammariens (Crustacea-Amphipoda) des îles Kerguelen et Crozet. Tethys 5: 635–708

Birstein YA, Vinogradov ME (1960) Pelagicheskie Gammaridy tropicheskoi chasti Tikhogo okeana (Pelagic Gammaridea from the tropical part of the Pacific Ocean). Trudy Inst Okeanol Akad Nauk SSSR 34: 165–241

Bogorov BG (1958) Biogeographical regions of the plankton of the North Western Pacific Ocean and their influence on the deep sea. Deep-Sea Res 5: 149–161

Bovallius C (1887) Contributions to a monograph of the Amphipoda Hyperiidea. Part I: 1. The families Tyronidae, Lanceolidae and Vibiliidae. K Sven Veenskapsakad Handl 21: 1–72, pls 1–10

Bovallius C (1889) Contributions to a monograph of the Amphipoda Hyperiidea. Part I: 2. The families Cyllopodidae, Paraphronimidae, Thaumatopsidae, Mimonectidae, Hyperiidae, Phronimidae and Anchylomeridae. K Sven Vetenskapsakad Handl 22: 1–434

Bowman TE (1960) The pelagic amphipod genus *Parathemisto* (Hyperiidea: Hyperiidae) in the North Pacific and adjacent Arctic Ocean. Proc US Nat Mus 112: 343–392

Bowman TE (1973) Pelagic amphipods of the genus *Hyperia* and closely related genera (Hyperiidea: Hyperiidae). Smithson Contrib Zool 136: 1–76

Bowman TE (1978) Revision of the pelagic amphipod genus *Primno* (Hyperiidea: Phrosinidae). Smithson Contrib Zool 275: 1–23

Bowman TE (1985) The correct identity of the pelagic amphipod *Primno macropa,* with a diagnosis of *Primno abyssalis* (Hyperiidea: Phrosinidae). Proc Biol Soc Wash 98: 121–126

Bowman TE, Meyers CD, Hicks SD (1963) Notes on associations between hyperiid amphipods and medusae in Chesapeake and Narragansett Bays and the Niantic River. Chesapeake Sci 4: 141–146

Chevreux E (1919) Revision des Scinidae provenant des campagnes de S.A.S. du le Prince de Monaco. Bull Inst Océanogr (Monaco) 352: 1–23

Conlan KE (1989) Delayed reproduction and adult dimorphism in males of the amphipod genus *Jassa* (Corophioidea: Ischyroceridae): an explanation for systematic confusion. J Crustacean Biol 9: 601–625

Conlan KE (1990) Revision of the crustacean amphipod genus *Jassa* Leach (Corophioidea: Ischyroceridae). Can J Zool 68: 2031–2075

Dana JD (1853) The Crustacea. Part II. US Explor (Philadelphia) 13: 687–1618. pls 1–96

De Broyer C (1983) Recherches sur la systématique et l'évolution des crustacés amphipodes gammarides antarctiques et subantarctiques. Thèse Doctorat en Sciences. Université de Louvain, 468 pp, 123 pls

De Broyer C, Jazdzewski K (1993) Contribution to the marine biodiversity inventory. A checklist of the Amphipoda (Crustacea) of the Southern Ocean. Doc Travail IRScNB 73: 1–154

Dick RI (1970) Hyperiidea (Crustacea: Amphipoda): keys to South African genera and species and a distribution list. Ann S Afr Mus 57: 25–86

Emison WB (1968) Feeding preferences of the Adélie penguin at Cape Crozier, Ross Island. Antarct Res Ser 12: 191–212

Guérin FE (1828) Uroptère. Encycl Méth Hist Nat 10: 345–832

Guérin-Méneville FE (1836) Description de quelques genres nouveaux appartenant à la famille des hypérines. Mag Zool 6: 1–10

Hurley DE (1955) Pelagic amphipods of the suborder Hyperiidea in New Zealand waters. I: Systematics. Trans R Soc NZ 83: 119–194

Hurley DE (1960) Amphipoda Hyperiidea. BANZ Antarct Res Exp Rep 8: 107–114

Hurley DE (1965) A common but hitherto undescribed species of *Orchomenella* (Crustacea Amphipoda: family Lysianassidae) from the Ross Sea. Trans R Soc NZ 6: 107–113

Hurley DE (1969) Amphipoda Hyperiidea. In: Antarctic map folio series, Folio 11, Distribution of selected groups of marine invertebrates in waters south of 35°S latitude. Am Geogr Soc: 32–34

Kane JE (1962) Amphipoda from waters south of New Zealand. NZ J Sci 5: 295–315

Kane JE (1966) The distribution of *Parathemisto gaudichaudii* (Guér.), with observations on its life history in the 0° to 20°E sector of the Southern Ocean. Discovery Rep 34: 163–198

Kröyer HN (1838) Grönlands Amfipoder beskriven af Henrik Kröyer. Vidensk Selskabs Naturvidensk Og Math Aphandlinger 7: 229–326

Littlepage JL, Pearse JS (1962) Biological and oceanographic observations under an Antarctic ice shelf. Science 137: 679–681

Lowry JK, Bullock S (1976) Catalogue of the marine gammaridean Amphipoda of the Southern Oceans. Bull R Soc NZ 16: 1–167

Ramirez FC, Viñas MD (1985) Hyperiid amphipods found in Argentine shelf waters. Physis (Buenos Aires) 43: 25–37

Raymont JEG (1983) Zooplankton. Plankton and productivity in the oceans. vol. 2. Pergamon Press, New York, pp 1–824

Sars GO (1883) Oversigt af Norges crustaceer med forelobige Bemaerkninger over de nye eller mindre bekjendte Arter. (Podophthalmata-Cumacea-Isopoda-Amphipoda) (med 6 autographiske Planker). Christiana Vidensk Forhandl 18, Fremlagt i Modetden 13 de Oktober

Sars GO (1895) An account of the Crustacea of Norway. 1. Amphipoda: 1–711

Schellenberg A (1926) Die Gammariden der deutschen Tiefsee-Expedition. Wiss Ergeb Dtsch Tiefsee Exp Valdivia 23: 195–243

Schellenberg A (1931) Gammariden und Caprelliden des Magellangebietes, Südgeorgiens und der Westantarktis. Further Zool Res Swed Ant Exp 1901–1903, 2: 1–290

Schneppenheim R, Weigmann-Haass R (1986) Morphological and electrophoretic studies of the genus *Themisto* (Amphipoda: Hyperiidea) from the South and North Atlantic. Polar Biol 6: 215–225

Semenova TN (1976) Systematics and distribution of pelagic amphipods of the family Vibiliidae (Hyperiidea) in the waters off New Zealand. Akad Nauk SSSR 105: 135–146

Senna A (1908) Su alcuni anfipodi iperini del plancton di Messina. Bull Soc Entomol Ital 38: 153–175

Sheader M (1975) Factors influencing change in the phenotype of the planktonic amphipod *Parathemisto gaudichaudii* (Guérin). J Mar Biol Assoc UK 55: 887–891

Sheader M, Evans F (1974) The taxonomic relationship of *Parathemisto gaudichaudii* (Guérin) and *P. gracilipes* (Norman), with a key to the genus *Parathemisto*. J Mar Biol Assoc UK 54: 915–924

Stebbing TRR (1888) Report on the Amphipoda collected by H.M.S. *Challenger* during the years 1873–76. Rep Sci Res Voy *Challenger* Lond 29: 1–1737

Thomson GM (1880) New species of crustacea from New Zealand. Ann Mag Nat Hist 6: 1–6

Thurston MH (1974) Crustacea Amphipoda from Graham Land and the Scotia Arc, collected by operation Tabarin and the Falkland Islands dependencies survey, 1944–59. Sci Rep Br Antarct Surv Lond 85: 1–89

Thurston MH (1976) The vertical distribution and diurnal migration of the crustacea Amphipoda collected during the SOND cruise, 1965, II. The Hyperiidea and general discussion. J Mar Biol Assoc UK 56: 383–470

Vinogradov ME (1966) Hyperiidea (Amphipoda) collected by the Soviet Antarctic expedition on M/V *OB* south of 40°S. In: Pavlovskii E. P. (ed) Biol Rep Sov Antarct Exped (1955–1958) 1: 1–32

Vinogradov ME (1970) New data on the amphipods Hyperiidea Physosomata of the northwestern part of the Pacific Ocean. (In Russian.) Trudy Instituta okeanologii. Akad Nauk SSSR 86: 382–400 English translation: Fauna of the Kurile – Kamchatka Trench and its environment, Israel Program for Scientific Translations, Jerusalem, 1972, 398–418

Vinogradov ME, Volkov AF, Semenova TN (1982) Amphipoda-Hyperiidea des Weltmeeres. Akad Nauk SSSR 132: 1–493

Wagler E (1926) Amphipoda 2: Scinidae. Wiss Ergeb Dtsch Tiefsee Exp *Valdivia* 20: 317–446

Walker AO (1906) Preliminary description of new species of Amphipoda from the *Discovery* Antarctic Expedition, 1902–1904. Ann Mag Nat Hist 17: 452–458

Weigmann-Haass R (1983) Hyperiidea (Amphipoda). In: Distribution of some zooplankton groups in the inner Weddell Sea. Ber Polarforsch 9: 4–35

Weigmann-Haass R (1989) Zur Taxonomie und Verbreitung der Gattung *Hyperiella* Bovallius 1887 im antarktischen Teil des Atlantik (Crustacea: Amphipoda: Hyperiidae). Senckenb Biol 69: 177–191

Zumelzu JM (1980) Anfipodos Hyperidos encontrados frente a la costa de Valparaiso. Aspectos taxonomicos. Inv Mar Valparaiso 8: 145–182

## Stations and Sample Data for amphipods During R/V *Cariboo* Cruise in the Straits of Magellan

*Cyphocaris faurei*

| Station number | Depth (m) | Ind·m⁻³ | Station number | Depth (m) | Ind·m⁻³ | Station number | Depth (m) | Ind·m⁻³ | Station number | Depth (m) | Ind·m⁻³ | Station number | Depth (m) | Ind·m⁻³ |
|---|---|---|---|---|---|---|---|---|---|---|---|---|---|---|
| 5 | 200–160 | 0.13 | 10 | 600–500 | 0.00 | 14 | 500–400 | 0.00 | 18 | 180–160 | 0.00 | 22 | 30–20 | 0.00 |
|  | 160–140 | 0.00 |  | 500–400 | 0.00 |  | 400–300 | 0.00 |  | 160–120 | 0.00 |  | 20–10 | 0.00 |
|  | 140–115 | 0.00 |  | 400–300 | 0.00 |  | 300–250 | 0.00 |  | 120–100 | 0.00 |  | 10–0 | 0.00 |
|  | 115–100 | 0.00 |  | 300–250 | 0.00 |  | 250–200 | 0.00 |  | 100–80 | 0.00 |  |  |  |
|  | 100–80 | 0.00 |  | 250–200 | 0.00 |  | 200–150 | 0.00 |  | 80–60 | 0.00 |  |  |  |
|  | 80–60 | 0.00 |  | 200–150 | 0.00 |  | 150–100 | 0.00 |  | 60–40 | 0.00 |  |  |  |
|  | 60–40 | 0.03 |  | 150–100 | 0.00 |  | 100–90 | 0.00 |  | 40–20 | 0.00 |  |  |  |
|  | 40–20 | 0.00 |  | 100–50 | 0.00 |  | 90–50 | 0.00 |  | 20–10 | 0.00 |  |  |  |
|  | 20–0 | 0.00 |  | 50–0 | 0.00 |  | 50–0 | 0.00 |  | 10–0 | 0.00 |  |  |  |
| 6 | 500–455 | 0.02 | 11 | 500–400 | 0.00 | 15 | 434–300 | 0.01 | 19 | 106–75 | 0.00 | 23 | 30–20 | 0.00 |
|  | 455–400 | 0.00 |  | 400–300 | 0.00 |  | 300–200 | 0.01 |  | 75–50 | 0.00 |  | 20–10 | 0.00 |
|  | 400–300 | 0.02 |  | 300–250 | 0.00 |  | 200–150 | 0.02 |  | 50–25 | 0.00 |  | 10–0 | 0.00 |
|  | 300–250 | 0.05 |  | 250–200 | 0.00 |  | 150–100 | 0.00 |  | 25–0 | 0.00 |  |  |  |
|  | 250–200 | 0.04 |  | 200–150 | 0.00 |  | 100–80 | 0.00 |  |  |  |  |  |  |
|  | 200–150 | 0.25 |  | 150–100 | 0.00 |  | 80–60 | 0.04 |  |  |  |  |  |  |
|  | 150–100 | 0.05 |  | 100–50 | 0.01 |  | 60–40 | 0.00 |  |  |  |  |  |  |
|  | 100–50 | 0.02 |  | 50–25 | 0.04 |  | 40–20 | 0.00 |  |  |  |  |  |  |
|  | 50–0 | 0.00 |  | 25–0 | 0.00 |  | 20–0 | 0.00 |  |  |  |  |  |  |
| 7 | 600–500 | 0.01 | 12 | 130–100 | 0.03 | 16 | 80–60 | 0.00 | 20 | 80–75 | 0.00 | 26 | 30–20 | 0.00 |
|  | 500–400 | 0.00 |  | 100–50 | 0.00 |  | 60–40 | 0.00 |  | 75–60 | 0.00 |  | 20–10 | 0.00 |
|  | 400–300 | 0.02 |  | 50–25 | 0.00 |  | 40–20 | 0.00 |  | 60–40 | 0.00 |  | 10–0 | 0.00 |
|  | 300–250 | 0.02 |  | 25–0 | 0.00 |  | 20–0 | 0.00 |  | 40–20 | 0.00 |  |  |  |
|  | 250–200 | 0.03 |  |  |  |  |  |  |  | 20–0 | 0.00 |  |  |  |
|  | 200–150 | 0.05 |  |  |  |  |  |  |  |  |  |  |  |  |
|  | 150–100 | 0.04 |  |  |  |  |  |  |  |  |  |  |  |  |
|  | 100–50 | 0.06 |  |  |  |  |  |  |  |  |  |  |  |  |
|  | 50–0 | 0.00 |  |  |  |  |  |  |  |  |  |  |  |  |
| 9 | 900–700 | 0.00 | 13 | 450–400 | 0.00 | 17 | 140–120 | 0.00 | 21 | 40–20 | 0.00 |  |  |  |
|  | 700–600 | 0.00 |  | 400–300 | 0.01 |  | 120–100 | 0.00 |  | 20–0 | 0.00 |  |  |  |
|  | 600–400 | 0.01 |  | 300–250 | 0.00 |  | 100–80 | 0.00 |  |  |  |  |  |  |
|  | 400–300 | 0.00 |  | 250–200 | 0.00 |  | 80–60 | 0.00 |  |  |  |  |  |  |
|  | 300–200 | 0.00 |  | 200–150 | 0.02 |  | 60–40 | 0.00 |  |  |  |  |  |  |
|  | 200–140 | 0.00 |  | 150–100 | 0.03 |  | 40–30 | 0.00 |  |  |  |  |  |  |
|  | 140–100 | 0.00 |  | 100–50 | 0.00 |  | 30–20 | 0.00 |  |  |  |  |  |  |
|  | 100–50 | 0.03 |  | 50–25 | 0.00 |  | 20–10 | 0.00 |  |  |  |  |  |  |
|  | 50–0 | 0.00 |  | 25–0 | 0.00 |  | 10–0 | 0.00 |  |  |  |  |  |  |

*Abyssorchomene plebs*

| Station number | Depth (m) | Ind·m⁻³ | Station number | Depth (m) | Ind·m⁻³ | Station number | Depth (m) | Ind·m⁻³ | Station number | Depth (m) | Ind·m⁻³ | Station number | Depth (m) | Ind·m⁻³ |
|---|---|---|---|---|---|---|---|---|---|---|---|---|---|---|
| 5 | 200–160 | 0.00 | 10 | 600–500 | 0.00 | 14 | 500–400 | 0.00 | 18 | 180–160 | 0.00 | 22 | 30–20 | 0.02 |
|  | 160–140 | 0.00 |  | 500–400 | 0.00 |  | 400–300 | 0.00 |  | 160–120 | 0.00 |  | 20–10 | 0.00 |
|  | 140–115 | 0.00 |  | 400–300 | 0.00 |  | 300–250 | 0.00 |  | 120–100 | 0.00 |  | 10–0 | 0.00 |
|  | 115–100 | 0.00 |  | 300–250 | 0.00 |  | 250–200 | 0.00 |  | 100–80 | 0.00 |  |  |  |
|  | 100–80 | 0.00 |  | 250–200 | 0.00 |  | 200–150 | 0.00 |  | 80–60 | 0.00 |  |  |  |
|  | 80–60 | 0.00 |  | 200–150 | 0.00 |  | 150–100 | 0.00 |  | 60–40 | 0.00 |  |  |  |
|  | 60–40 | 0.00 |  | 150–100 | 0.00 |  | 100–90 | 0.00 |  | 40–20 | 0.00 |  |  |  |
|  | 40–20 | 0.00 |  | 100–50 | 0.00 |  | 90–50 | 0.00 |  | 20–10 | 0.00 |  |  |  |
|  | 20–0 | 0.00 |  | 50–0 | 0.00 |  | 50–0 | 0.00 |  | 10–0 | 0.00 |  |  |  |
| 6 | 500–455 | 0.00 | 11 | 500–400 | 0.01 | 15 | 434–300 | 0.00 | 19 | 106–75 | 0.00 | 23 | 30–20 | 0.02 |
|  | 455–400 | 0.00 |  | 400–300 | 0.00 |  | 300–200 | 0.00 |  | 75–50 | 0.00 |  | 20–10 | 0.00 |
|  | 400–300 | 0.00 |  | 300–250 | 0.00 |  | 200–150 | 0.00 |  | 50–25 | 0.00 |  | 10–0 | 0.00 |
|  | 300–250 | 0.00 |  | 250–200 | 0.00 |  | 150–100 | 0.00 |  | 25–0 | 0.00 |  |  |  |
|  | 250–200 | 0.00 |  | 200–150 | 0.00 |  | 100–80 | 0.00 |  |  |  |  |  |  |
|  | 200–150 | 0.00 |  | 150–100 | 0.00 |  | 80–60 | 0.00 |  |  |  |  |  |  |
|  | 150–100 | 0.00 |  | 100–50 | 0.00 |  | 60–40 | 0.00 |  |  |  |  |  |  |
|  | 100–50 | 0.00 |  | 50–25 | 0.00 |  | 40–20 | 0.00 |  |  |  |  |  |  |
|  | 50–0 | 0.00 |  | 25–0 | 0.00 |  | 20–0 | 0.00 |  |  |  |  |  |  |
| 7 | 600–500 | 0.01 | 12 | 130–100 | 0.00 | 16 | 80–60 | 0.00 | 20 | 80–75 | 0.00 | 26 | 30–20 | 0.00 |
|  | 500–400 | 0.01 |  | 100–50 | 0.00 |  | 60–40 | 0.00 |  | 75–60 | 0.00 |  | 20–10 | 0.00 |
|  | 400–300 | 0.00 |  | 50–25 | 0.00 |  | 40–20 | 0.00 |  | 60–40 | 0.00 |  | 10–0 | 0.00 |
|  | 300–250 | 0.00 |  | 25–0 | 0.00 |  | 20–0 | 0.00 |  | 40–20 | 0.00 |  |  |  |
|  | 250–200 | 0.00 |  |  |  |  |  |  |  | 20–0 | 0.00 |  |  |  |
|  | 200–150 | 0.00 |  |  |  |  |  |  |  |  |  |  |  |  |
|  | 150–100 | 0.00 |  |  |  |  |  |  |  |  |  |  |  |  |
|  | 100–50 | 0.00 |  |  |  |  |  |  |  |  |  |  |  |  |
|  | 50–0 | 0.00 |  |  |  |  |  |  |  |  |  |  |  |  |
| 9 | 900–700 | 0.00 | 13 | 450–400 | 0.00 | 17 | 140–120 | 0.00 | 21 | 40–20 | 0.00 |  |  |  |
|  | 700–600 | 0.00 |  | 400–300 | 0.00 |  | 120–100 | 0.00 |  | 20–0 | 0.00 |  |  |  |
|  | 600–400 | 0.00 |  | 300–250 | 0.00 |  | 100–80 | 0.00 |  |  |  |  |  |  |
|  | 400–300 | 0.00 |  | 250–200 | 0.00 |  | 80–60 | 0.00 |  |  |  |  |  |  |
|  | 300–200 | 0.00 |  | 200–150 | 0.00 |  | 60–40 | 0.00 |  |  |  |  |  |  |
|  | 200–140 | 0.00 |  | 150–100 | 0.00 |  | 40–30 | 0.00 |  |  |  |  |  |  |
|  | 140–100 | 0.00 |  | 100–50 | 0.00 |  | 30–20 | 0.00 |  |  |  |  |  |  |
|  | 100–50 | 0.00 |  | 50–25 | 0.00 |  | 20–10 | 0.00 |  |  |  |  |  |  |
|  | 50–0 | 0.00 |  | 25–0 | 0.00 |  | 10–0 | 0.00 |  |  |  |  |  |  |

*Thryphosella schellenbergi*

| Station number | Depth (m) | Ind·m$^{-3}$ | Station number | Depth (m) | Ind·m$^{-3}$ | Station number | Depth (m) | Ind·m$^{-3}$ | Station number | Depth (m) | Ind·m$^{-3}$ | Station number | Depth (m) | Ind·m$^{-3}$ |
|---|---|---|---|---|---|---|---|---|---|---|---|---|---|---|
| 5 | 200–160 | 0.00 | 10 | 600–500 | 0.00 | 14 | 500–400 | 0.00 | 18 | 180–160 | 0.00 | 22 | 30–20 | 0.00 |
|   | 160–140 | 0.00 |    | 500–400 | 0.00 |    | 400–300 | 0.00 |    | 160–120 | 0.00 |    | 20–10 | 0.00 |
|   | 140–115 | 0.00 |    | 400–300 | 0.00 |    | 300–250 | 0.00 |    | 120–100 | 0.00 |    | 10–0 | 0.00 |
|   | 115–100 | 0.00 |    | 300–250 | 0.00 |    | 250–200 | 0.00 |    | 100–80 | 0.00 |    |   |   |
|   | 100–80 | 0.00 |    | 250–200 | 0.00 |    | 200–150 | 0.00 |    | 80–60 | 0.00 |    |   |   |
|   | 80–60 | 0.00 |    | 200–150 | 0.00 |    | 150–100 | 0.00 |    | 60–40 | 0.00 |    |   |   |
|   | 60–40 | 0.00 |    | 150–100 | 0.00 |    | 100–90 | 0.00 |    | 40–20 | 0.00 |    |   |   |
|   | 40–20 | 0.00 |    | 100–50 | 0.00 |    | 90–50 | 0.00 |    | 20–10 | 0.00 |    |   |   |
|   | 20–0 | 0.00 |    | 50–0 | 0.00 |    | 50–0 | 0.00 |    | 10–0 | 0.00 |    |   |   |
| 6 | 500–455 | 0.00 | 11 | 500–400 | 0.00 | 15 | 434–300 | 0.00 | 19 | 106–75 | 0.00 | 23 | 30–20 | 0.02 |
|   | 455–400 | 0.00 |    | 400–300 | 0.00 |    | 300–200 | 0.00 |    | 75–50 | 0.00 |    | 20–10 | 0.02 |
|   | 400–300 | 0.00 |    | 300–250 | 0.00 |    | 200–150 | 0.00 |    | 50–25 | 0.00 |    | 10–0 | 0.00 |
|   | 300–250 | 0.00 |    | 250–200 | 0.00 |    | 150–100 | 0.00 |    | 25–0 | 0.00 |    |   |   |
|   | 250–200 | 0.00 |    | 200–150 | 0.00 |    | 100–80 | 0.00 |    |   |   |    |   |   |
|   | 200–150 | 0.00 |    | 150–100 | 0.00 |    | 80–60 | 0.00 |    |   |   |    |   |   |
|   | 150–100 | 0.00 |    | 100–50 | 0.00 |    | 60–40 | 0.00 |    |   |   |    |   |   |
|   | 100–50 | 0.00 |    | 50–25 | 0.00 |    | 40–20 | 0.00 |    |   |   |    |   |   |
|   | 50–0 | 0.00 |    | 25–0 | 0.00 |    | 20–0 | 0.00 |    |   |   |    |   |   |
| 7 | 600–500 | 0.00 | 12 | 130–100 | 0.00 | 16 | 80–60 | 0.00 | 20 | 80–75 | 0.00 | 26 | 30–20 | 0.00 |
|   | 500–400 | 0.00 |    | 100–50 | 0.00 |    | 60–40 | 0.00 |    | 75–60 | 0.00 |    | 20–10 | 0.00 |
|   | 400–300 | 0.00 |    | 50–25 | 0.00 |    | 40–20 | 0.00 |    | 60–40 | 0.00 |    | 10–0 | 0.00 |
|   | 300–250 | 0.00 |    | 25–0 | 0.00 |    | 20–0 | 0.00 |    | 40–20 | 0.00 |    |   |   |
|   | 250–200 | 0.00 |    |   |   |    |   |   |    | 20–0 | 0.00 |    |   |   |
|   | 200–150 | 0.00 |    |   |   |    |   |   |    |   |   |    |   |   |
|   | 150–100 | 0.00 |    |   |   |    |   |   |    |   |   |    |   |   |
|   | 100–50 | 0.00 |    |   |   |    |   |   |    |   |   |    |   |   |
|   | 50–0 | 0.00 |    |   |   |    |   |   |    |   |   |    |   |   |
| 9 | 900–700 | 0.00 | 13 | 450–400 | 0.00 | 17 | 140–120 | 0.00 | 21 | 40–20 | 0.01 |   |   |   |
|   | 700–600 | 0.00 |    | 400–300 | 0.00 |    | 120–100 | 0.00 |    | 20–0 | 0.01 |   |   |   |
|   | 600–400 | 0.00 |    | 300–250 | 0.00 |    | 100–80 | 0.00 |    |   |   |   |   |   |
|   | 400–300 | 0.00 |    | 250–200 | 0.00 |    | 80–60 | 0.00 |    |   |   |   |   |   |
|   | 300–200 | 0.00 |    | 200–150 | 0.00 |    | 60–40 | 0.00 |    |   |   |   |   |   |
|   | 200–140 | 0.00 |    | 150–100 | 0.00 |    | 40–30 | 0.00 |    |   |   |   |   |   |
|   | 140–100 | 0.00 |    | 100–50 | 0.00 |    | 30–20 | 0.00 |    |   |   |   |   |   |
|   | 100–50 | 0.00 |    | 50–25 | 0.00 |    | 20–10 | 0.00 |    |   |   |   |   |   |
|   | 50–0 | 0.00 |    | 25–0 | 0.00 |    | 10–0 | 0.00 |    |   |   |   |   |   |

*Eusirus antarcticus*

| Station number | Depth (m) | Ind·m⁻³ | Station number | Depth (m) | Ind·m⁻³ | Station number | Depth (m) | Ind·m⁻³ | Station number | Depth (m) | Ind·m⁻³ | Station number | Depth (m) | Ind·m⁻³ |
|---|---|---|---|---|---|---|---|---|---|---|---|---|---|---|
| 5 | 200–160 | 0.00 | 10 | 600–500 | 0.00 | 14 | 500–400 | 0.00 | 18 | 180–160 | 0.00 | 22 | 30–20 | 0.00 |
|  | 160–140 | 0.00 |  | 500–400 | 0.00 |  | 400–300 | 0.00 |  | 160–120 | 0.00 |  | 20–10 | 0.00 |
|  | 140–115 | 0.00 |  | 400–300 | 0.00 |  | 300–250 | 0.00 |  | 120–100 | 0.00 |  | 10–0 | 0.00 |
|  | 115–100 | 0.00 |  | 300–250 | 0.00 |  | 250–200 | 0.00 |  | 100–80 | 0.00 |  |  |  |
|  | 100–80 | 0.00 |  | 250–200 | 0.00 |  | 200–150 | 0.00 |  | 80–60 | 0.00 |  |  |  |
|  | 80–60 | 0.00 |  | 200–150 | 0.00 |  | 150–100 | 0.00 |  | 60–40 | 0.00 |  |  |  |
|  | 60–40 | 0.00 |  | 150–100 | 0.00 |  | 100–90 | 0.00 |  | 40–20 | 0.00 |  |  |  |
|  | 40–20 | 0.00 |  | 100–50 | 0.00 |  | 90–50 | 0.00 |  | 20–10 | 0.00 |  |  |  |
|  | 20–0 | 0.00 |  | 50–0 | 0.00 |  | 50–0 | 0.00 |  | 10–0 | 0.00 |  |  |  |
| 6 | 500–455 | 0.00 | 11 | 500–400 | 0.00 | 15 | 434–300 | 0.00 | 19 | 106–75 | 0.00 | 23 | 30–20 | 0.00 |
|  | 455–400 | 0.00 |  | 400–300 | 0.00 |  | 300–200 | 0.00 |  | 75–50 | 0.00 |  | 20–10 | 0.00 |
|  | 400–300 | 0.00 |  | 300–250 | 0.00 |  | 200–150 | 0.00 |  | 50–25 | 0.00 |  | 10–0 | 0.00 |
|  | 300–250 | 0.00 |  | 250–200 | 0.00 |  | 150–100 | 0.00 |  | 25–0 | 0.00 |  |  |  |
|  | 250–200 | 0.00 |  | 200–150 | 0.00 |  | 100–80 | 0.00 |  |  |  |  |  |  |
|  | 200–150 | 0.00 |  | 150–100 | 0.00 |  | 80–60 | 0.00 |  |  |  |  |  |  |
|  | 150–100 | 0.00 |  | 100–50 | 0.00 |  | 60–40 | 0.00 |  |  |  |  |  |  |
|  | 100–50 | 0.00 |  | 50–25 | 0.00 |  | 40–20 | 0.00 |  |  |  |  |  |  |
|  | 50–0 | 0.00 |  | 25–0 | 0.00 |  | 20–0 | 0.00 |  |  |  |  |  |  |
| 7 | 600–500 | 0.00 | 12 | 130–100 | 0.00 | 16 | 80–60 | 0.00 | 20 | 80–75 | 0.00 | 26 | 30–20 | 0.00 |
|  | 500–400 | 0.00 |  | 100–50 | 0.00 |  | 60–40 | 0.00 |  | 75–60 | 0.00 |  | 20–10 | 0.00 |
|  | 400–300 | 0.00 |  | 50–25 | 0.00 |  | 40–20 | 0.00 |  | 60–40 | 0.00 |  | 10–0 | 0.00 |
|  | 300–250 | 0.00 |  | 25–0 | 0.00 |  | 20–0 | 0.00 |  | 40–20 | 0.00 |  |  |  |
|  | 250–200 | 0.00 |  |  |  |  |  |  |  | 20–0 | 0.00 |  |  |  |
|  | 200–150 | 0.00 |  |  |  |  |  |  |  |  |  |  |  |  |
|  | 150–100 | 0.00 |  |  |  |  |  |  |  |  |  |  |  |  |
|  | 100–50 | 0.00 |  |  |  |  |  |  |  |  |  |  |  |  |
|  | 50–0 | 0.00 |  |  |  |  |  |  |  |  |  |  |  |  |
| 9 | 900–700 | 0.00 | 13 | 450–400 | 0.00 | 17 | 140–120 | 0.00 | 21 | 40–20 | 0.01 |  |  |  |
|  | 700–600 | 0.00 |  | 400–300 | 0.00 |  | 120–100 | 0.00 |  | 20–0 | 0.01 |  |  |  |
|  | 600–400 | 0.00 |  | 300–250 | 0.00 |  | 100–80 | 0.00 |  |  |  |  |  |  |
|  | 400–300 | 0.00 |  | 250–200 | 0.00 |  | 80–60 | 0.00 |  |  |  |  |  |  |
|  | 300–200 | 0.00 |  | 200–150 | 0.00 |  | 60–40 | 0.00 |  |  |  |  |  |  |
|  | 200–140 | 0.00 |  | 150–100 | 0.00 |  | 40–30 | 0.00 |  |  |  |  |  |  |
|  | 140–100 | 0.00 |  | 100–50 | 0.00 |  | 30–20 | 0.00 |  |  |  |  |  |  |
|  | 100–50 | 0.00 |  | 50–25 | 0.00 |  | 20–10 | 0.00 |  |  |  |  |  |  |
|  | 50–0 | 0.00 |  | 25–0 | 0.00 |  | 10–0 | 0.00 |  |  |  |  |  |  |

*Rhachotropis schellenbergi*

| Station number | Depth (m) | Ind·m⁻³ | Station number | Depth (m) | Ind·m⁻³ | Station number | Depth (m) | Ind·m⁻³ | Station number | Depth (m) | Ind·m⁻³ | Station number | Depth (m) | Ind·m⁻³ |
|---|---|---|---|---|---|---|---|---|---|---|---|---|---|---|
| 5 | 200–160 | 0.00 | 10 | 600–500 | 0.00 | 14 | 500–400 | 0.00 | 18 | 180–160 | 0.00 | 22 | 30–20 | 0.00 |
|  | 160–140 | 0.00 |  | 500–400 | 0.00 |  | 400–300 | 0.00 |  | 160–120 | 0.00 |  | 20–10 | 0.00 |
|  | 140–115 | 0.00 |  | 400–300 | 0.00 |  | 300–250 | 0.00 |  | 120–100 | 0.00 |  | 10–0 | 0.00 |
|  | 115–100 | 0.00 |  | 300–250 | 0.00 |  | 250–200 | 0.00 |  | 100–80 | 0.00 |  |  |  |
|  | 100–80 | 0.00 |  | 250–200 | 0.00 |  | 200–150 | 0.00 |  | 80–60 | 0.00 |  |  |  |
|  | 80–60 | 0.00 |  | 200–150 | 0.00 |  | 150–100 | 0.00 |  | 60–40 | 0.00 |  |  |  |
|  | 60–40 | 0.00 |  | 150–100 | 0.00 |  | 100–90 | 0.00 |  | 40–20 | 0.00 |  |  |  |
|  | 40–20 | 0.00 |  | 100–50 | 0.00 |  | 90–50 | 0.00 |  | 20–10 | 0.00 |  |  |  |
|  | 20–0 | 0.00 |  | 50–0 | 0.00 |  | 50–0 | 0.00 |  | 10–0 | 0.00 |  |  |  |
| 6 | 500–455 | 0.00 | 11 | 500–400 | 0.00 | 15 | 434–300 | 0.00 | 19 | 106–75 | 0.07 | 23 | 30–20 | 0.00 |
|  | 455–400 | 0.00 |  | 400–300 | 0.00 |  | 300–200 | 0.00 |  | 75–50 | 0.01 |  | 20–10 | 0.00 |
|  | 400–300 | 0.00 |  | 300–250 | 0.00 |  | 200–150 | 0.02 |  | 50–25 | 0.00 |  | 10–0 | 0.00 |
|  | 300–250 | 0.00 |  | 250–200 | 0.00 |  | 150–100 | 0.00 |  | 25–0 | 0.00 |  |  |  |
|  | 250–200 | 0.00 |  | 200–150 | 0.00 |  | 100–80 | 0.00 |  |  |  |  |  |  |
|  | 200–150 | 0.00 |  | 150–100 | 0.00 |  | 80–60 | 0.00 |  |  |  |  |  |  |
|  | 150–100 | 0.00 |  | 100–50 | 0.00 |  | 60–40 | 0.00 |  |  |  |  |  |  |
|  | 100–50 | 0.00 |  | 50–25 | 0.00 |  | 40–20 | 0.00 |  |  |  |  |  |  |
|  | 50–0 | 0.00 |  | 25–0 | 0.00 |  | 20–0 | 0.00 |  |  |  |  |  |  |
| 7 | 600–500 | 0.00 | 12 | 130–100 | 0.00 | 16 | 80–60 | 0.00 | 20 | 80–75 | 0.00 | 26 | 30–20 | 0.00 |
|  | 500–400 | 0.00 |  | 100–50 | 0.00 |  | 60–40 | 0.00 |  | 75–60 | 0.00 |  | 20–10 | 0.00 |
|  | 400–300 | 0.00 |  | 50–25 | 0.00 |  | 40–20 | 0.00 |  | 60–40 | 0.00 |  | 10–0 | 0.00 |
|  | 300–250 | 0.00 |  | 25–0 | 0.00 |  | 20–0 | 0.00 |  | 40–20 | 0.00 |  |  |  |
|  | 250–200 | 0.00 |  |  |  |  |  |  |  | 20–0 | 0.00 |  |  |  |
|  | 200–150 | 0.00 |  |  |  |  |  |  |  |  |  |  |  |  |
|  | 150–100 | 0.00 |  |  |  |  |  |  |  |  |  |  |  |  |
|  | 100–50 | 0.00 |  |  |  |  |  |  |  |  |  |  |  |  |
|  | 50–0 | 0.00 |  |  |  |  |  |  |  |  |  |  |  |  |
| 9 | 900–700 | 0.00 | 13 | 450–400 | 0.00 | 17 | 140–120 | 0.00 | 21 | 40–20 | 0.00 |  |  |  |
|  | 700–600 | 0.00 |  | 400–300 | 0.00 |  | 120–100 | 0.00 |  | 20–0 | 0.00 |  |  |  |
|  | 600–400 | 0.00 |  | 300–250 | 0.00 |  | 100–80 | 0.00 |  |  |  |  |  |  |
|  | 400–300 | 0.00 |  | 250–200 | 0.00 |  | 80–60 | 0.00 |  |  |  |  |  |  |
|  | 300–200 | 0.00 |  | 200–150 | 0.00 |  | 60–40 | 0.00 |  |  |  |  |  |  |
|  | 200–140 | 0.00 |  | 150–100 | 0.00 |  | 40–30 | 0.00 |  |  |  |  |  |  |
|  | 140–100 | 0.00 |  | 100–50 | 0.00 |  | 30–20 | 0.00 |  |  |  |  |  |  |
|  | 100–50 | 0.00 |  | 50–25 | 0.00 |  | 20–10 | 0.00 |  |  |  |  |  |  |
|  | 50–0 | 0.00 |  | 25–0 | 0.00 |  | 10–0 | 0.00 |  |  |  |  |  |  |

*Jassa alonsoae*

| Station number | Depth (m) | Ind·m⁻³ | Station number | Depth (m) | Ind·m⁻³ | Station number | Depth (m) | Ind·m⁻³ | Station number | Depth (m) | Ind·m⁻³ | Station number | Depth (m) | Ind·m⁻³ |
|---|---|---|---|---|---|---|---|---|---|---|---|---|---|---|
| 5 | 200–160 | 0.00 | 10 | 600–500 | 0.00 | 14 | 500–400 | 0.00 | 18 | 180–160 | 0.00 | 22 | 30–20 | 0.00 |
| | 160–140 | 0.00 | | 500–400 | 0.00 | | 400–300 | 0.00 | | 160–120 | 0.00 | | 20–10 | 0.02 |
| | 140–115 | 0.00 | | 400–300 | 0.00 | | 300–250 | 0.00 | | 120–100 | 0.00 | | 10–0 | 0.00 |
| | 115–100 | 0.00 | | 300–250 | 0.00 | | 250–200 | 0.00 | | 100–80 | 0.00 | | | |
| | 100–80 | 0.00 | | 250–200 | 0.00 | | 200–150 | 0.00 | | 80–60 | 0.00 | | | |
| | 80–60 | 0.00 | | 200–150 | 0.00 | | 150–100 | 0.00 | | 60–40 | 0.00 | | | |
| | 60–40 | 0.00 | | 150–100 | 0.00 | | 100–90 | 0.00 | | 40–20 | 0.00 | | | |
| | 40–20 | 0.00 | | 100–50 | 0.00 | | 90–50 | 0.00 | | 20–10 | 0.00 | | | |
| | 20–0 | 0.00 | | 50–0 | 0.00 | | 50–0 | 0.00 | | 10–0 | 0.00 | | | |
| 6 | 500–455 | 0.00 | 11 | 500–400 | 0.00 | 15 | 434–300 | 0.00 | 19 | 106–75 | 0.00 | 23 | 30–20 | 0.00 |
| | 455–400 | 0.00 | | 400–300 | 0.00 | | 300–200 | 0.00 | | 75–50 | 0.00 | | 20–10 | 0.00 |
| | 400–300 | 0.00 | | 300–250 | 0.00 | | 200–150 | 0.00 | | 50–25 | 0.00 | | 10–0 | 0.00 |
| | 300–250 | 0.00 | | 250–200 | 0.00 | | 150–100 | 0.00 | | 25–0 | 0.00 | | | |
| | 250–200 | 0.00 | | 200–150 | 0.00 | | 100–80 | 0.00 | | | | | | |
| | 200–150 | 0.00 | | 150–100 | 0.00 | | 80–60 | 0.00 | | | | | | |
| | 150–100 | 0.00 | | 100–50 | 0.00 | | 60–40 | 0.00 | | | | | | |
| | 100–50 | 0.00 | | 50–25 | 0.00 | | 40–20 | 0.00 | | | | | | |
| | 50–0 | 0.00 | | 25–0 | 0.00 | | 20–0 | 0.00 | | | | | | |
| 7 | 600–500 | 0.00 | 12 | 130–100 | 0.00 | 16 | 80–60 | 0.00 | 20 | 80–75 | 0.00 | 26 | 30–20 | 0.00 |
| | 500–400 | 0.00 | | 100–50 | 0.00 | | 60–40 | 0.00 | | 75–60 | 0.00 | | 20–10 | 0.00 |
| | 400–300 | 0.00 | | 50–25 | 0.00 | | 40–20 | 0.00 | | 60–40 | 0.00 | | 10–0 | 0.00 |
| | 300–250 | 0.00 | | 25–0 | 0.00 | | 20–0 | 0.00 | | 40–20 | 0.00 | | | |
| | 250–200 | 0.00 | | | | | | | | 20–0 | 0.00 | | | |
| | 200–150 | 0.00 | | | | | | | | | | | | |
| | 150–100 | 0.00 | | | | | | | | | | | | |
| | 100–50 | 0.00 | | | | | | | | | | | | |
| | 50–0 | 0.00 | | | | | | | | | | | | |
| 9 | 900–700 | 0.00 | 13 | 450–400 | 0.00 | 17 | 140–120 | 0.00 | 21 | 40–20 | 0.00 | | | |
| | 700–600 | 0.00 | | 400–300 | 0.00 | | 120–100 | 0.00 | | 20–0 | 0.00 | | | |
| | 600–400 | 0.00 | | 300–250 | 0.00 | | 100–80 | 0.00 | | | | | | |
| | 400–300 | 0.00 | | 250–200 | 0.00 | | 80–60 | 0.00 | | | | | | |
| | 300–200 | 0.00 | | 200–150 | 0.00 | | 60–40 | 0.00 | | | | | | |
| | 200–140 | 0.00 | | 150–100 | 0.00 | | 40–30 | 0.00 | | | | | | |
| | 140–100 | 0.00 | | 100–50 | 0.00 | | 30–20 | 0.00 | | | | | | |
| | 100–50 | 0.00 | | 50–25 | 0.00 | | 20–10 | 0.00 | | | | | | |
| | 50–0 | 0.00 | | 25–0 | 0.00 | | 10–0 | 0.00 | | | | | | |

*Scina borealis*

| Station number | Depth (m) | Ind·m⁻³ | Station number | Depth (m) | Ind·m⁻³ | Station number | Depth (m) | Ind·m⁻³ | Station number | Depth (m) | Ind·m⁻³ | Station number | Depth (m) | Ind·m⁻³ |
|---|---|---|---|---|---|---|---|---|---|---|---|---|---|---|
| 5 | 200–160 | 0.00 | 10 | 600–500 | 0.00 | 14 | 500–400 | 0.00 | 18 | 180–160 | 0.00 | 22 | 30–20 | 0.00 |
|  | 160–140 | 0.00 |  | 500–400 | 0.00 |  | 400–300 | 0.00 |  | 160–120 | 0.00 |  | 20–10 | 0.00 |
|  | 140–115 | 0.00 |  | 400–300 | 0.00 |  | 300–250 | 0.00 |  | 120–100 | 0.00 |  | 10–0 | 0.00 |
|  | 115–100 | 0.00 |  | 300–250 | 0.00 |  | 250–200 | 0.00 |  | 100–80 | 0.00 |  |  |  |
|  | 100–80 | 0.00 |  | 250–200 | 0.00 |  | 200–150 | 0.00 |  | 80–60 | 0.00 |  |  |  |
|  | 80–60 | 0.00 |  | 200–150 | 0.00 |  | 150–100 | 0.00 |  | 60–40 | 0.00 |  |  |  |
|  | 60–40 | 0.00 |  | 150–100 | 0.00 |  | 100–90 | 0.00 |  | 40–20 | 0.00 |  |  |  |
|  | 40–20 | 0.00 |  | 100–50 | 0.00 |  | 90–50 | 0.00 |  | 20–10 | 0.00 |  |  |  |
|  | 20–0 | 0.00 |  | 50–0 | 0.00 |  | 50–0 | 0.00 |  | 10–0 | 0.00 |  |  |  |
| 6 | 500–455 | 0.00 | 11 | 500–400 | 0.00 | 15 | 434–300 | 0.00 | 19 | 106–75 | 0.00 | 23 | 30–20 | 0.00 |
|  | 455–400 | 0.00 |  | 400–300 | 0.01 |  | 300–200 | 0.00 |  | 75–50 | 0.00 |  | 20–10 | 0.00 |
|  | 400–300 | 0.00 |  | 300–250 | 0.00 |  | 200–150 | 0.00 |  | 50–25 | 0.00 |  | 10–0 | 0.00 |
|  | 300–250 | 0.00 |  | 250–200 | 0.00 |  | 150–100 | 0.00 |  | 25–0 | 0.00 |  |  |  |
|  | 250–200 | 0.00 |  | 200–150 | 0.01 |  | 100–80 | 0.00 |  |  |  |  |  |  |
|  | 200–150 | 0.00 |  | 150–100 | 0.00 |  | 80–60 | 0.00 |  |  |  |  |  |  |
|  | 150–100 | 0.00 |  | 100–50 | 0.00 |  | 60–40 | 0.00 |  |  |  |  |  |  |
|  | 100–50 | 0.00 |  | 50–25 | 0.00 |  | 40–20 | 0.00 |  |  |  |  |  |  |
|  | 50–0 | 0.00 |  | 25–0 | 0.00 |  | 20–0 | 0.00 |  |  |  |  |  |  |
| 7 | 600–500 | 0.00 | 12 | 130–100 | 0.00 | 16 | 80–60 | 0.00 | 20 | 80–75 | 0.00 | 26 | 30–20 | 0.00 |
|  | 500–400 | 0.00 |  | 100–50 | 0.00 |  | 60–40 | 0.00 |  | 75–60 | 0.00 |  | 20–10 | 0.00 |
|  | 400–300 | 0.00 |  | 50–25 | 0.00 |  | 40–20 | 0.00 |  | 60–40 | 0.00 |  | 10–0 | 0.00 |
|  | 300–250 | 0.00 |  | 25–0 | 0.00 |  | 20–0 | 0.00 |  | 40–20 | 0.00 |  |  |  |
|  | 250–200 | 0.00 |  |  |  |  |  |  |  | 20–0 | 0.00 |  |  |  |
|  | 200–150 | 0.00 |  |  |  |  |  |  |  |  |  |  |  |  |
|  | 150–100 | 0.00 |  |  |  |  |  |  |  |  |  |  |  |  |
|  | 100–50 | 0.00 |  |  |  |  |  |  |  |  |  |  |  |  |
|  | 50–0 | 0.00 |  |  |  |  |  |  |  |  |  |  |  |  |
| 9 | 900–700 | 0.00 | 13 | 450–400 | 0.00 | 17 | 140–120 | 0.00 | 21 | 40–20 | 0.00 |  |  |  |
|  | 700–600 | 0.01 |  | 400–300 | 0.00 |  | 120–100 | 0.00 |  | 20–0 | 0.00 |  |  |  |
|  | 600–400 | 0.01 |  | 300–250 | 0.00 |  | 100–80 | 0.00 |  |  |  |  |  |  |
|  | 400–300 | 0.00 |  | 250–200 | 0.00 |  | 80–60 | 0.00 |  |  |  |  |  |  |
|  | 300–200 | 0.00 |  | 200–150 | 0.00 |  | 60–40 | 0.00 |  |  |  |  |  |  |
|  | 200–140 | 0.10 |  | 150–100 | 0.00 |  | 40–30 | 0.00 |  |  |  |  |  |  |
|  | 140–100 | 0.00 |  | 100–50 | 0.00 |  | 30–20 | 0.00 |  |  |  |  |  |  |
|  | 100–50 | 0.03 |  | 50–25 | 0.00 |  | 20–10 | 0.00 |  |  |  |  |  |  |
|  | 50–0 | 0.00 |  | 25–0 | 0.00 |  | 10–0 | 0.00 |  |  |  |  |  |  |

*Cyllopus magellanicus*

| Station number | Depth (m) | Ind·m$^{-3}$ | Station number | Depth (m) | Ind·m$^{-3}$ | Station number | Depth (m) | Ind·m$^{-3}$ | Station number | Depth (m) | Ind·m$^{-3}$ | Station number | Depth (m) | Ind·m$^{-3}$ |
|---|---|---|---|---|---|---|---|---|---|---|---|---|---|---|
| 5 | 200–160 | 0.02 | 10 | 600–500 | 0.00 | 14 | 500–400 | 0.00 | 18 | 180–160 | 0.00 | 22 | 30–20 | 0.00 |
|   | 160–140 | 0.00 |   | 500–400 | 0.00 |   | 400–300 | 0.00 |   | 160–120 | 0.00 |   | 20–10 | 0.00 |
|   | 140–115 | 0.02 |   | 400–300 | 0.00 |   | 300–250 | 0.00 |   | 120–100 | 0.00 |   | 10–0 | 0.00 |
|   | 115–100 | 0.00 |   | 300–250 | 0.00 |   | 250–200 | 0.00 |   | 100–80 | 0.00 |   |   |   |
|   | 100–80 | 0.00 |   | 250–200 | 0.00 |   | 200–150 | 0.00 |   | 80–60 | 0.00 |   |   |   |
|   | 80–60 | 0.00 |   | 200–150 | 0.00 |   | 150–100 | 0.00 |   | 60–40 | 0.00 |   |   |   |
|   | 60–40 | 0.00 |   | 150–100 | 0.00 |   | 100–90 | 0.00 |   | 40–20 | 0.00 |   |   |   |
|   | 40–20 | 0.00 |   | 100–50 | 0.00 |   | 90–50 | 0.00 |   | 20–10 | 0.00 |   |   |   |
|   | 20–0 | 0.00 |   | 50–0 | 0.00 |   | 50–0 | 0.00 |   | 10–0 | 0.00 |   |   |   |
| 6 | 500–455 | 0.00 | 11 | 500–400 | 0.00 | 15 | 434–300 | 0.00 | 19 | 106–75 | 0.00 | 23 | 30–20 | 0.00 |
|   | 455–400 | 0.00 |   | 400–300 | 0.00 |   | 300–200 | 0.00 |   | 75–50 | 0.00 |   | 20–10 | 0.00 |
|   | 400–300 | 0.00 |   | 300–250 | 0.00 |   | 200–150 | 0.00 |   | 50–25 | 0.00 |   | 10–0 | 0.00 |
|   | 300–250 | 0.00 |   | 250–200 | 0.00 |   | 150–100 | 0.00 |   | 25–0 | 0.00 |   |   |   |
|   | 250–200 | 0.00 |   | 200–150 | 0.00 |   | 100–80 | 0.00 |   |   |   |   |   |   |
|   | 200–150 | 0.00 |   | 150–100 | 0.00 |   | 80–60 | 0.00 |   |   |   |   |   |   |
|   | 150–100 | 0.00 |   | 100–50 | 0.00 |   | 60–40 | 0.00 |   |   |   |   |   |   |
|   | 100–50 | 0.00 |   | 50–25 | 0.00 |   | 40–20 | 0.00 |   |   |   |   |   |   |
|   | 50–0 | 0.00 |   | 25–0 | 0.00 |   | 20–0 | 0.00 |   |   |   |   |   |   |
| 7 | 600–500 | 0.00 | 12 | 130–100 | 0.00 | 16 | 80–60 | 0.00 | 20 | 80–75 | 0.00 | 26 | 30–20 | 0.00 |
|   | 500–400 | 0.00 |   | 100–50 | 0.00 |   | 60–40 | 0.00 |   | 75–60 | 0.00 |   | 20–10 | 0.00 |
|   | 400–300 | 0.00 |   | 50–25 | 0.00 |   | 40–20 | 0.00 |   | 60–40 | 0.00 |   | 10–0 | 0.00 |
|   | 300–250 | 0.00 |   | 25–0 | 0.00 |   | 20–0 | 0.00 |   | 40–20 | 0.00 |   |   |   |
|   | 250–200 | 0.00 |   |   |   |   |   |   |   | 20–0 | 0.00 |   |   |   |
|   | 200–150 | 0.00 |   |   |   |   |   |   |   |   |   |   |   |   |
|   | 150–100 | 0.00 |   |   |   |   |   |   |   |   |   |   |   |   |
|   | 100–50 | 0.00 |   |   |   |   |   |   |   |   |   |   |   |   |
|   | 50–0 | 0.00 |   |   |   |   |   |   |   |   |   |   |   |   |
| 9 | 900–700 | 0.00 | 13 | 450–400 | 0.00 | 17 | 140–120 | 0.00 | 21 | 40–20 | 0.00 |   |   |   |
|   | 700–600 | 0.00 |   | 400–300 | 0.00 |   | 120–100 | 0.00 |   | 20–0 | 0.00 |   |   |   |
|   | 600–400 | 0.00 |   | 300–250 | 0.00 |   | 100–80 | 0.00 |   |   |   |   |   |   |
|   | 400–300 | 0.00 |   | 250–200 | 0.00 |   | 80–60 | 0.00 |   |   |   |   |   |   |
|   | 300–200 | 0.00 |   | 200–150 | 0.00 |   | 60–40 | 0.00 |   |   |   |   |   |   |
|   | 200–140 | 0.00 |   | 150–100 | 0.00 |   | 40–30 | 0.00 |   |   |   |   |   |   |
|   | 140–100 | 0.00 |   | 100–50 | 0.00 |   | 30–20 | 0.00 |   |   |   |   |   |   |
|   | 100–50 | 0.00 |   | 50–25 | 0.00 |   | 20–10 | 0.00 |   |   |   |   |   |   |
|   | 50–0 | 0.00 |   | 25–0 | 0.00 |   | 10–0 | 0.00 |   |   |   |   |   |   |

*Hyperiella dilatata*

| Station number | Depth (m) | Ind·m⁻³ | Station number | Depth (m) | Ind·m⁻³ | Station number | Depth (m) | Ind·m⁻³ | Station number | Depth (m) | Ind·m⁻³ | Station number | Depth (m) | Ind·m⁻³ | Station number | Depth (m) | Ind·m⁻³ |
|---|---|---|---|---|---|---|---|---|---|---|---|---|---|---|---|---|---|
| 5 | 200–160 | 0.00 | 10 | 600–500 | 0.00 | 14 | 500–400 | 0.00 | 18 | 180–160 | 0.00 | 22 | 30–20 | 0.00 | | | |
| | 160–140 | 0.00 | | 500–400 | 0.00 | | 400–300 | 0.00 | | 160–120 | 0.00 | | 20–10 | 0.00 | | | |
| | 140–115 | 0.00 | | 400–300 | 0.00 | | 300–250 | 0.00 | | 120–100 | 0.00 | | 10–0 | 0.00 | | | |
| | 115–100 | 0.00 | | 300–250 | 0.00 | | 250–200 | 0.00 | | 100–80 | 0.00 | | | | | | |
| | 100–80 | 0.00 | | 250–200 | 0.00 | | 200–150 | 0.00 | | 80–60 | 0.00 | | | | | | |
| | 80–60 | 0.03 | | 200–150 | 0.00 | | 150–100 | 0.00 | | 60–40 | 0.00 | | | | | | |
| | 60–40 | 0.00 | | 150–100 | 0.00 | | 100–90 | 0.00 | | 40–20 | 0.00 | | | | | | |
| | 40–20 | 0.00 | | 100–50 | 0.00 | | 90–50 | 0.00 | | 20–10 | 0.00 | | | | | | |
| | 20–0 | 0.00 | | 50–0 | 0.00 | | 50–0 | 0.00 | | 10–0 | 0.00 | | | | | | |
| 6 | 500–455 | 0.00 | 11 | 500–400 | 0.00 | 15 | 434–300 | 0.00 | 19 | 106–75 | 0.00 | 23 | 30–20 | 0.00 | | | |
| | 455–400 | 0.00 | | 400–300 | 0.00 | | 300–200 | 0.00 | | 75–50 | 0.00 | | 20–10 | 0.00 | | | |
| | 400–300 | 0.00 | | 300–250 | 0.00 | | 200–150 | 0.00 | | 50–25 | 0.00 | | 10–0 | 0.00 | | | |
| | 300–250 | 0.00 | | 250–200 | 0.00 | | 150–100 | 0.00 | | 25–0 | 0.00 | | | | | | |
| | 250–200 | 0.00 | | 200–150 | 0.00 | | 100–80 | 0.00 | | | | | | | | | |
| | 200–150 | 0.00 | | 150–100 | 0.00 | | 80–60 | 0.00 | | | | | | | | | |
| | 150–100 | 0.00 | | 100–50 | 0.00 | | 60–40 | 0.00 | | | | | | | | | |
| | 100–50 | 0.00 | | 50–25 | 0.00 | | 40–20 | 0.00 | | | | | | | | | |
| | 50–0 | 0.00 | | 25–0 | 0.00 | | 20–0 | 0.00 | | | | | | | | | |
| 7 | 600–500 | 0.00 | 12 | 130–100 | 0.00 | 16 | 80–60 | 0.00 | 20 | 80–75 | 0.00 | 26 | 30–20 | 0.00 | | | |
| | 500–400 | 0.00 | | 100–50 | 0.00 | | 60–40 | 0.00 | | 75–60 | 0.00 | | 20–10 | 0.00 | | | |
| | 400–300 | 0.00 | | 50–25 | 0.00 | | 40–20 | 0.00 | | 60–40 | 0.00 | | 10–0 | 0.00 | | | |
| | 300–250 | 0.00 | | 25–0 | 0.00 | | 20–0 | 0.00 | | 40–20 | 0.00 | | | | | | |
| | 250–200 | 0.00 | | | | | | | | 20–0 | 0.00 | | | | | | |
| | 200–150 | 0.00 | | | | | | | | | | | | | | | |
| | 150–100 | 0.00 | | | | | | | | | | | | | | | |
| | 100–50 | 0.00 | | | | | | | | | | | | | | | |
| | 50–0 | 0.00 | | | | | | | | | | | | | | | |
| 9 | 900–700 | 0.00 | 13 | 450–400 | 0.00 | 17 | 140–120 | 0.00 | 21 | 40–20 | 0.00 | | | | | | |
| | 700–600 | 0.00 | | 400–300 | 0.00 | | 120–100 | 0.00 | | 20–0 | 0.00 | | | | | | |
| | 600–400 | 0.00 | | 300–250 | 0.00 | | 100–80 | 0.00 | | | | | | | | | |
| | 400–300 | 0.00 | | 250–200 | 0.00 | | 80–60 | 0.00 | | | | | | | | | |
| | 300–200 | 0.00 | | 200–150 | 0.00 | | 60–40 | 0.00 | | | | | | | | | |
| | 200–140 | 0.00 | | 150–100 | 0.00 | | 40–30 | 0.00 | | | | | | | | | |
| | 140–100 | 0.00 | | 100–50 | 0.00 | | 30–20 | 0.00 | | | | | | | | | |
| | 100–50 | 0.00 | | 50–25 | 0.00 | | 20–10 | 0.00 | | | | | | | | | |
| | 50–0 | 0.00 | | 25–0 | 0.00 | | 10–0 | 0.00 | | | | | | | | | |

*Hyperoche mediterranea*

| Station number | Depth (m) | Ind·m⁻³ | Station number | Depth (m) | Ind·m⁻³ | Station number | Depth (m) | Ind·m⁻³ | Station number | Depth (m) | Ind·m⁻³ | Station number | Depth (m) | Ind·m⁻³ |
|---|---|---|---|---|---|---|---|---|---|---|---|---|---|---|
| 5 | 200–160 | 0.00 | 10 | 600–500 | 0.00 | 14 | 500–400 | 0.00 | 18 | 180–160 | 0.00 | 22 | 30–20 | 0.00 |
|   | 160–140 | 0.00 |   | 500–400 | 0.00 |   | 400–300 | 0.00 |   | 160–120 | 0.00 |   | 20–10 | 0.00 |
|   | 140–115 | 0.00 |   | 400–300 | 0.00 |   | 300–250 | 0.00 |   | 120–100 | 0.00 |   | 10–0 | 0.00 |
|   | 115–100 | 0.00 |   | 300–250 | 0.00 |   | 250–200 | 0.00 |   | 100–80 | 0.00 |   |   |   |
|   | 100–80 | 0.00 |   | 250–200 | 0.00 |   | 200–150 | 0.00 |   | 80–60 | 0.00 |   |   |   |
|   | 80–60 | 0.00 |   | 200–150 | 0.00 |   | 150–100 | 0.00 |   | 60–40 | 0.00 |   |   |   |
|   | 60–40 | 0.00 |   | 150–100 | 0.00 |   | 100–90 | 0.00 |   | 40–20 | 0.00 |   |   |   |
|   | 40–20 | 0.00 |   | 100–50 | 0.00 |   | 90–50 | 0.00 |   | 20–10 | 0.00 |   |   |   |
|   | 20–0 | 0.03 |   | 50–0 | 0.00 |   | 50–0 | 0.00 |   | 10–0 | 0.00 |   |   |   |
| 6 | 500–455 | 0.00 | 11 | 500–400 | 0.00 | 15 | 434–300 | 0.00 | 19 | 106–75 | 0.00 | 23 | 30–20 | 0.00 |
|   | 455–400 | 0.00 |   | 400–300 | 0.00 |   | 300–200 | 0.00 |   | 75–50 | 0.00 |   | 20–10 | 0.00 |
|   | 400–300 | 0.00 |   | 300–250 | 0.00 |   | 200–150 | 0.00 |   | 50–25 | 0.00 |   | 10–0 | 0.00 |
|   | 300–250 | 0.00 |   | 250–200 | 0.00 |   | 150–100 | 0.00 |   | 25–0 | 0.00 |   |   |   |
|   | 250–200 | 0.00 |   | 200–150 | 0.00 |   | 100–80 | 0.00 |   |   |   |   |   |   |
|   | 200–150 | 0.00 |   | 150–100 | 0.00 |   | 80–60 | 0.00 |   |   |   |   |   |   |
|   | 150–100 | 0.00 |   | 100–50 | 0.00 |   | 60–40 | 0.00 |   |   |   |   |   |   |
|   | 100–50 | 0.00 |   | 50–25 | 0.00 |   | 40–20 | 0.00 |   |   |   |   |   |   |
|   | 50–0 | 0.00 |   | 25–0 | 0.00 |   | 20–0 | 0.00 |   |   |   |   |   |   |
| 7 | 600–500 | 0.00 | 12 | 130–100 | 0.00 | 16 | 80–60 | 0.00 | 20 | 80–75 | 0.00 | 26 | 30–20 | 0.00 |
|   | 500–400 | 0.00 |   | 100–50 | 0.00 |   | 60–40 | 0.00 |   | 75–60 | 0.00 |   | 20–10 | 0.00 |
|   | 400–300 | 0.00 |   | 50–25 | 0.00 |   | 40–20 | 0.00 |   | 60–40 | 0.00 |   | 10–0 | 0.00 |
|   | 300–250 | 0.00 |   | 25–0 | 0.00 |   | 20–0 | 0.00 |   | 40–20 | 0.00 |   |   |   |
|   | 250–200 | 0.00 |   |   |   |   |   |   |   | 20–0 | 0.00 |   |   |   |
|   | 200–150 | 0.00 |   |   |   |   |   |   |   |   |   |   |   |   |
|   | 150–100 | 0.00 |   |   |   |   |   |   |   |   |   |   |   |   |
|   | 100–50 | 0.00 |   |   |   |   |   |   |   |   |   |   |   |   |
|   | 50–0 | 0.00 |   |   |   |   |   |   |   |   |   |   |   |   |
| 9 | 900–700 | 0.00 | 13 | 450–400 | 0.00 | 17 | 140–120 | 0.00 | 21 | 40–20 | 0.00 |   |   |   |
|   | 700–600 | 0.00 |   | 400–300 | 0.00 |   | 120–100 | 0.00 |   | 20–0 | 0.00 |   |   |   |
|   | 600–400 | 0.00 |   | 300–250 | 0.00 |   | 100–80 | 0.00 |   |   |   |   |   |   |
|   | 400–300 | 0.00 |   | 250–200 | 0.00 |   | 80–60 | 0.00 |   |   |   |   |   |   |
|   | 300–200 | 0.00 |   | 200–150 | 0.00 |   | 60–40 | 0.00 |   |   |   |   |   |   |
|   | 200–140 | 0.00 |   | 150–100 | 0.00 |   | 40–30 | 0.00 |   |   |   |   |   |   |
|   | 140–100 | 0.00 |   | 100–50 | 0.00 |   | 30–20 | 0.00 |   |   |   |   |   |   |
|   | 100–50 | 0.00 |   | 50–25 | 0.00 |   | 20–10 | 0.00 |   |   |   |   |   |   |
|   | 50–0 | 0.00 |   | 25–0 | 0.00 |   | 10–0 | 0.00 |   |   |   |   |   |   |

*Hyperoche medusarum*

| Station number | Depth (m) | Ind·m$^{-3}$ | Station number | Depth (m) | Ind·m$^{-3}$ | Station number | Depth (m) | Ind·m$^{-3}$ | Station number | Depth (m) | Ind·m$^{-3}$ | Station number | Depth (m) | Ind·m$^{-3}$ |
|---|---|---|---|---|---|---|---|---|---|---|---|---|---|---|
| 5 | 200–160 | 0.00 | 10 | 600–500 | 0.00 | 14 | 500–400 | 0.02 | 18 | 180–160 | 0.02 | 22 | 30–20 | 0.00 |
|   | 160–140 | 0.00 |    | 500–400 | 0.00 |    | 400–300 | 0.00 |    | 160–120 | 0.00 |    | 20–10 | 0.00 |
|   | 140–115 | 0.00 |    | 400–300 | 0.00 |    | 300–250 | 0.01 |    | 120–100 | 0.00 |    | 10–0 | 0.00 |
|   | 115–100 | 0.00 |    | 300–250 | 0.00 |    | 250–200 | 0.00 |    | 100–80 | 0.00 |    |   |   |
|   | 100–80 | 0.00 |    | 250–200 | 0.00 |    | 200–150 | 0.02 |    | 80–60 | 0.00 |    |   |   |
|   | 80–60 | 0.00 |    | 200–150 | 0.00 |    | 150–100 | 0.00 |    | 60–40 | 0.00 |    |   |   |
|   | 60–40 | 0.00 |    | 150–100 | 0.00 |    | 100–90 | 0.00 |    | 40–20 | 0.00 |    |   |   |
|   | 40–20 | 0.00 |    | 100–50 | 0.00 |    | 90–50 | 0.00 |    | 20–10 | 0.00 |    |   |   |
|   | 20–0 | 0.00 |    | 50–0 | 0.00 |    | 50–0 | 0.00 |    | 10–0 | 0.00 |    |   |   |
| 6 | 500–455 | 0.00 | 11 | 500–400 | 0.00 | 15 | 434–300 | 0.00 | 19 | 106–75 | 0.00 | 23 | 30–20 | 0.00 |
|   | 455–400 | 0.00 |    | 400–300 | 0.00 |    | 300–200 | 0.00 |    | 75–50 | 0.00 |    | 20–10 | 0.00 |
|   | 400–300 | 0.00 |    | 300–250 | 0.00 |    | 200–150 | 0.00 |    | 50–25 | 0.00 |    | 10–0 | 0.00 |
|   | 300–250 | 0.00 |    | 250–200 | 0.00 |    | 150–100 | 0.00 |    | 25–0 | 0.00 |    |   |   |
|   | 250–200 | 0.00 |    | 200–150 | 0.00 |    | 100–80 | 0.00 |    |   |   |    |   |   |
|   | 200–150 | 0.00 |    | 150–100 | 0.00 |    | 80–60 | 0.00 |    |   |   |    |   |   |
|   | 150–100 | 0.00 |    | 100–50 | 0.00 |    | 60–40 | 0.00 |    |   |   |    |   |   |
|   | 100–50 | 0.00 |    | 50–25 | 0.00 |    | 40–20 | 0.00 |    |   |   |    |   |   |
|   | 50–0 | 0.00 |    | 25–0 | 0.00 |    | 20–0 | 0.00 |    |   |   |    |   |   |
| 7 | 600–500 | 0.00 | 12 | 130–100 | 0.00 | 16 | 80–60 | 0.00 | 20 | 80–75 | 0.00 | 26 | 30–20 | 0.00 |
|   | 500–400 | 0.00 |    | 100–50 | 0.00 |    | 60–40 | 0.00 |    | 75–60 | 0.00 |    | 20–10 | 0.00 |
|   | 400–300 | 0.01 |    | 50–25 | 0.00 |    | 40–20 | 0.00 |    | 60–40 | 0.00 |    | 10–0 | 0.00 |
|   | 300–250 | 0.00 |    | 25–0 | 0.00 |    | 20–0 | 0.00 |    | 40–20 | 0.00 |    |   |   |
|   | 250–200 | 0.00 |    |   |   |    |   |   |    | 20–0 | 0.00 |    |   |   |
|   | 200–150 | 0.00 |    |   |   |    |   |   |    |   |   |    |   |   |
|   | 150–100 | 0.00 |    |   |   |    |   |   |    |   |   |    |   |   |
|   | 100–50 | 0.00 |    |   |   |    |   |   |    |   |   |    |   |   |
|   | 50–0 | 0.00 |    |   |   |    |   |   |    |   |   |    |   |   |
| 9 | 900–700 | 0.00 | 13 | 450–400 | 0.00 | 17 | 140–120 | 0.00 | 21 | 40–20 | 0.00 |   |   |   |
|   | 700–600 | 0.00 |    | 400–300 | 0.00 |    | 120–100 | 0.00 |    | 20–0 | 0.00 |    |   |   |
|   | 600–400 | 0.00 |    | 300–250 | 0.00 |    | 100–80 | 0.00 |    |   |   |    |   |   |
|   | 400–300 | 0.00 |    | 250–200 | 0.02 |    | 80–60 | 0.00 |    |   |   |    |   |   |
|   | 300–200 | 0.00 |    | 200–150 | 0.02 |    | 60–40 | 0.00 |    |   |   |    |   |   |
|   | 200–140 | 0.00 |    | 150–100 | 0.00 |    | 40–30 | 0.00 |    |   |   |    |   |   |
|   | 140–100 | 0.00 |    | 100–50 | 0.00 |    | 30–20 | 0.00 |    |   |   |    |   |   |
|   | 100–50 | 0.00 |    | 50–25 | 0.00 |    | 20–10 | 0.00 |    |   |   |    |   |   |
|   | 50–0 | 0.00 |    | 25–0 | 0.00 |    | 10–0 | 0.00 |    |   |   |    |   |   |

*Themisto gaudichaudii*

| Station number | Depth (m) | Ind·m$^{-3}$ | Station number | Depth (m) | Ind·m$^{-3}$ | Station number | Depth (m) | Ind·m$^{-3}$ | Station number | Depth (m) | Ind·m$^{-3}$ | Station number | Depth (m) | Ind·m$^{-3}$ |
|---|---|---|---|---|---|---|---|---|---|---|---|---|---|---|
| 5 | 200–160 | 0.05 | 10 | 600–500 | 0.00 | 14 | 500–400 | 0.05 | 18 | 180–160 | 0.39 | 22 | 30–20 | 0.41 |
|   | 160–140 | 0.04 |   | 500–400 | 0.00 |   | 400–300 | 0.00 |   | 160–120 | 0.55 |   | 20–10 | 0.47 |
|   | 140–115 | 0.00 |   | 400–300 | 0.00 |   | 300–250 | 0.00 |   | 120–100 | 0.03 |   | 10–0 | 0.17 |
|   | 115–100 | 0.07 |   | 300–250 | 0.01 |   | 250–200 | 0.07 |   | 100–80 | 0.21 |   |   |   |
|   | 100–80 | 0.04 |   | 250–200 | 0.00 |   | 200–150 | 0.35 |   | 80–60 | 0.00 |   |   |   |
|   | 80–60 | 0.00 |   | 200–150 | 0.00 |   | 150–100 | 0.27 |   | 60–40 | 0.18 |   |   |   |
|   | 60–40 | 0.00 |   | 150–100 | 0.51 |   | 100–90 | 0.00 |   | 40–20 | 0.09 |   |   |   |
|   | 40–20 | 0.00 |   | 100–50 | 1.39 |   | 90–50 | 0.04 |   | 20–10 | 0.71 |   |   |   |
|   | 20–0 | 0.03 |   | 50–0 | 0.58 |   | 50–0 | 0.26 |   | 10–0 | 0.26 |   |   |   |
| 6 | 500–455 | 0.00 | 11 | 500–400 | 0.01 | 15 | 434–300 | 0.00 | 19 | 106–75 | 0.25 | 23 | 30–20 | 1.19 |
|   | 455–400 | 0.00 |   | 400–300 | 0.01 |   | 300–200 | 0.07 |   | 75–50 | 0.45 |   | 20–10 | 2.08 |
|   | 400–300 | 0.00 |   | 300–250 | 0.00 |   | 200–150 | 0.02 |   | 50–25 | 0.68 |   | 10–0 | 0.87 |
|   | 300–250 | 0.00 |   | 250–200 | 0.00 |   | 150–100 | 0.16 |   | 25–0 | 0.22 |   |   |   |
|   | 250–200 | 0.00 |   | 200–150 | 0.03 |   | 100–80 | 0.08 |   |   |   |   |   |   |
|   | 200–150 | 0.03 |   | 150–100 | 0.09 |   | 80–60 | 0.35 |   |   |   |   |   |   |
|   | 150–100 | 0.00 |   | 100–50 | 0.29 |   | 60–40 | 0.17 |   |   |   |   |   |   |
|   | 100–50 | 0.01 |   | 50–25 | 1.40 |   | 40–20 | 0.61 |   |   |   |   |   |   |
|   | 50–0 | 0.54 |   | 25–0 | 2.98 |   | 20–0 | 0.27 |   |   |   |   |   |   |
| 7 | 600–500 | 0.00 | 12 | 130–100 | 0.26 | 16 | 80–60 | 0.37 | 20 | 80–75 | 0.35 | 26 | 30–20 | 11.02 |
|   | 500–400 | 0.00 |   | 100–50 | 0.25 |   | 60–40 | 0.45 |   | 75–60 | 0.18 |   | 20–10 | 2.08 |
|   | 400–300 | 0.33 |   | 50–25 | 1.09 |   | 40–20 | 1.61 |   | 60–40 | 0.26 |   | 10–0 | 2.56 |
|   | 300–250 | 0.00 |   | 25–0 | 1.86 |   | 20–0 | 1.45 |   | 40–20 | 1.94 |   |   |   |
|   | 250–200 | 0.01 |   |   |   |   |   |   |   | 20–0 | 1.08 |   |   |   |
|   | 200–150 | 0.07 |   |   |   |   |   |   |   |   |   |   |   |   |
|   | 150–100 | 0.00 |   |   |   |   |   |   |   |   |   |   |   |   |
|   | 100–50 | 0.06 |   |   |   |   |   |   |   |   |   |   |   |   |
|   | 50–0 | 0.54 |   |   |   |   |   |   |   |   |   |   |   |   |
| 9 | 900–700 | 0.03 | 13 | 450–400 | 0.01 | 17 | 140–120 | 0.33 | 21 | 40–20 | 1.85 |   |   |   |
|   | 700–600 | 0.01 |   | 400–300 | 0.00 |   | 120–100 | 0.03 |   | 20–0 | 2.10 |   |   |   |
|   | 600–400 | 0.00 |   | 300–250 | 0.00 |   | 100–80 | 0.13 |   |   |   |   |   |   |
|   | 400–300 | 0.00 |   | 250–200 | 0.26 |   | 80–60 | 1.20 |   |   |   |   |   |   |
|   | 300–200 | 0.00 |   | 200–150 | 0.28 |   | 60–40 | 0.85 |   |   |   |   |   |   |
|   | 200–140 | 0.10 |   | 150–100 | 0.08 |   | 40–30 | 2.40 |   |   |   |   |   |   |
|   | 140–100 | 0.00 |   | 100–50 | 0.00 |   | 30–20 | 1.66 |   |   |   |   |   |   |
|   | 100–50 | 0.03 |   | 50–25 | 1.36 |   | 20–10 | 0.98 |   |   |   |   |   |   |
|   | 50–0 | 0.13 |   | 25–0 | 2.88 |   | 10–0 | 0.05 |   |   |   |   |   |   |

*Primno macropa*

| Station number | Depth (m) | Ind·m⁻³ | Station number | Depth (m) | Ind·m⁻³ | Station number | Depth (m) | Ind·m⁻³ | Station number | Depth (m) | Ind·m⁻³ | Station number | Depth (m) | Ind·m⁻³ |
|---|---|---|---|---|---|---|---|---|---|---|---|---|---|---|
| 5 | 200–160 | 0.10 | 10 | 600–500 | 0.00 | 14 | 500–400 | 0.00 | 18 | 180–160 | 0.00 | 22 | 30–20 | 0.00 |
|  | 160–140 | 0.00 |  | 500–400 | 0.00 |  | 400–300 | 0.00 |  | 160–120 | 0.00 |  | 20–10 | 0.00 |
|  | 140–115 | 0.00 |  | 400–300 | 0.00 |  | 300–250 | 0.00 |  | 120–100 | 0.00 |  | 10–0 | 0.00 |
|  | 115–100 | 0.04 |  | 300–250 | 0.00 |  | 250–200 | 0.00 |  | 100–80 | 0.00 |  |  |  |
|  | 100–80 | 0.00 |  | 250–200 | 0.00 |  | 200–150 | 0.00 |  | 80–60 | 0.00 |  |  |  |
|  | 80–60 | 0.05 |  | 200–150 | 0.00 |  | 150–100 | 0.00 |  | 60–40 | 0.00 |  |  |  |
|  | 60–40 | 0.00 |  | 150–100 | 0.00 |  | 100–90 | 0.00 |  | 40–20 | 0.00 |  |  |  |
|  | 40–20 | 0.00 |  | 100–50 | 0.00 |  | 90–50 | 0.00 |  | 20–10 | 0.00 |  |  |  |
|  | 20–0 | 0.06 |  | 50–0 | 0.00 |  | 50–0 | 0.00 |  | 10–0 | 0.00 |  |  |  |
| 6 | 500–455 | 0.00 | 11 | 500–400 | 0.00 | 15 | 434–300 | 0.00 | 19 | 106–75 | 0.00 | 23 | 30–20 | 0.00 |
|  | 455–400 | 0.01 |  | 400–300 | 0.01 |  | 300–200 | 0.00 |  | 75–50 | 0.00 |  | 20–10 | 0.00 |
|  | 400–300 | 0.01 |  | 300–250 | 0.00 |  | 200–150 | 0.00 |  | 50–25 | 0.00 |  | 10–0 | 0.00 |
|  | 300–250 | 0.00 |  | 250–200 | 0.00 |  | 150–100 | 0.00 |  | 25–0 | 0.00 |  |  |  |
|  | 250–200 | 0.02 |  | 200–150 | 0.00 |  | 100–80 | 0.00 |  |  |  |  |  |  |
|  | 200–150 | 0.00 |  | 150–100 | 0.01 |  | 80–60 | 0.00 |  |  |  |  |  |  |
|  | 150–100 | 0.02 |  | 100–50 | 0.00 |  | 60–40 | 0.00 |  |  |  |  |  |  |
|  | 100–50 | 0.02 |  | 50–25 | 0.00 |  | 40–20 | 0.00 |  |  |  |  |  |  |
|  | 50–0 | 0.02 |  | 25–0 | 0.00 |  | 20–0 | 0.00 |  |  |  |  |  |  |
| 7 | 600–500 | 0.00 | 12 | 130–100 | 0.00 | 16 | 80–60 | 0.00 | 20 | 80–75 | 0.00 | 26 | 30–20 | 0.00 |
|  | 500–400 | 0.01 |  | 100–50 | 0.00 |  | 60–40 | 0.00 |  | 75–60 | 0.00 |  | 20–10 | 0.00 |
|  | 400–300 | 0.01 |  | 50–25 | 0.00 |  | 40–20 | 0.00 |  | 60–40 | 0.00 |  | 10–0 | 0.00 |
|  | 300–250 | 0.00 |  | 25–0 | 0.00 |  | 20–0 | 0.00 |  | 40–20 | 0.00 |  |  |  |
|  | 250–200 | 0.04 |  |  |  |  |  |  |  | 20–0 | 0.00 |  |  |  |
|  | 200–150 | 0.00 |  |  |  |  |  |  |  |  |  |  |  |  |
|  | 150–100 | 0.00 |  |  |  |  |  |  |  |  |  |  |  |  |
|  | 100–50 | 0.00 |  |  |  |  |  |  |  |  |  |  |  |  |
|  | 50–0 | 0.02 |  |  |  |  |  |  |  |  |  |  |  |  |
| 9 | 900–700 | 0.00 | 13 | 450–400 | 0.00 | 17 | 140–120 | 0.00 | 21 | 40–20 | 0.00 |  |  |  |
|  | 700–600 | 0.00 |  | 400–300 | 0.00 |  | 120–100 | 0.00 |  | 20–0 | 0.00 |  |  |  |
|  | 600–400 | 0.00 |  | 300–250 | 0.00 |  | 100–80 | 0.00 |  |  |  |  |  |  |
|  | 400–300 | 0.00 |  | 250–200 | 0.00 |  | 80–60 | 0.00 |  |  |  |  |  |  |
|  | 300–200 | 0.00 |  | 200–150 | 0.00 |  | 60–40 | 0.00 |  |  |  |  |  |  |
|  | 200–140 | 0.00 |  | 150–100 | 0.00 |  | 40–30 | 0.00 |  |  |  |  |  |  |
|  | 140–100 | 0.00 |  | 100–50 | 0.00 |  | 30–20 | 0.00 |  |  |  |  |  |  |
|  | 100–50 | 0.00 |  | 50–25 | 0.00 |  | 20–10 | 0.00 |  |  |  |  |  |  |
|  | 50–0 | 0.00 |  | 25–0 | 0.00 |  | 10–0 | 0.00 |  |  |  |  |  |  |

*Chapter 4* **Euphausiids**

L. Guglielmo, T. Antezana, N. Crescenti, and A. Granata

**Systematic Account**

# Subphylum CRUSTACEA
## Class MALACOSTRACA
### Order EUPHAUSIACEA

**Family Euphausiidae**

## 4.1 *Euphausia vallentini* Stebbing, 1900

*Euphausia vallentini* Stebbing, 1900: p. 545, pl. 37 – Holt and Tattersall, 1906: p. 3 – Tattersall, 1908: pp. 13, 14, pl. 4 figs. 4–6 – Hansen, 1911: p. 30 – Hansen, 1913: p. 32, pl. 5 fig. 1a–f – Zimmer, 1914: p. 427 – Zimmer, 1915: pp. 178, 179 – John, 1936: pp. 211–214, figs. 12-14, 30b – Sheard, 1953: tabs. 2,3,9 – Boden, 1954: p. 215, fig. 15g–j – Bary, 1956: pp. 442, 443, fig. 11 – Mauchline and Fischer, 1969: p. 64, fig. 20 – Mayer, 1969: pp. 252–255, figs. 2a, 6–8 – Ramírez, 1971: pp. 395–397, pl. 6 fig. 4, pl. 10 – Ramirez, 1973: p. 11 – Antezana et al., 1976: pp. 60, 62, figs. 2–9b, 11b – Antezana and Brinton, 1981: p. 688, figs. 216–233 – Antezana, 1985: pp. 84, 85.
*Euphausia splendens* (partim) Sars, 1885: p. 80, pl. 13 figs. 7–17.
*Euphausia patachonica* Colosi, 1917: p. 187, pl. 14, figs.11–14, pl. 15 figs. 15–20.

Body length –
    females: 17.3–22.5 mm (100 specimens)
    males: 18.2–23.3 mm (50 specimens)

## Description

Lower margin of carapace provided with one lateral denticle. Rostrum short and pointed, triangular in shape, and as long or longer than wide. Frontal plate not well differentiated. Eyes spherical and of medium size. Basal segment of antennular peduncle bearing large, wide and rounded lobe or lappet on inner distal margin; lappet projected forward and somewhat upward. In large specimens, 2nd segment bearing small denticles on inner and outer distal corners. Third segment bearing high dorsal keel with rounded upper margin and straight frontal margin. Antennal scale extending beyond distal margin of 2nd antennular segment. Lateral spine of basiscerite reaching distal margin of eye. Third abdominal segment projected on distal dorsal margin in short and fragile accuminated process. This process may be completely lacking in small and large specimens. Preanal spine provided with one or two secondary spines, which are larger in females than in males. Subterminal spines of telson lacking denticles on inner margins. However, in some cases, they may be present but in small numbers. Terminal process of petasma slender. Proximal heel poorly developed. Distal end bifurcated with inner ramous expanded and provided, in some cases, with small denticle. Outer ramous as long as inner one.

Proximal process longer and more robust than terminal process and expanding, on distal end, in 2 membranous leaf-like lappets. Anterior lappet straight and as long as posterior one which bears small denticle on inner side. Wide secondary spine situated next to bifurcation of lappets. Lateral process strongly curved and bearing strong tooth on distal end and minute denticle-like process. Setiferous lobe bearing 4 seate on inner margin, 2 on outer margin and 2 at tip. Coxal plates developed laterally in 2 finger-shaped prominences. Anteriorly, there is a small opening of the pocket; this is delimited laterally by 2 folds of the coxal plates and posteriorly by a rather triangular-shaped ridge of the sternite (Guglielmo and Costanzo 1978).

## Remarks

The diagnostic features of *E. vallentini* vary to such an extent, that they resemble *E. lucens,* another sub-Antarctic species also found in the northern range of the Chilean fjord region (Antezana 1976). In fact, Mayer (1969), based on individuals collected off the South American continent, proposed that both species be included as one, with a wide range of individual variation of the petasma and of the first segment of the antennule. Mayer's illustrations, however, showed the characteristics allowing for the identification of both species. Furthermore, the author's figs. 4, 5, 7 and 8 correspond to juveniles. Mayer's questioning the validity of both species, particularly the frequent absence of a spine on the 3rd abdominal segment of *E. vallentini*, and variability and overlapping of other diagnostic features, allows us to point out the following as additional distinguishing features: size and shape of the dorsal keel of the 3rd antennular segment, denticles on the inner border of the subterminal spines, and distribution of setae on the setiferous lobe of the petasma. It should be remarked that specimens collected in the northern extension of the Chilean fjords and in the Humboldt Current up to 38°S (Antezana 1976, 1978, 1981) fit the above description of the species; none carry a spine on the 3rd abdominal segment.

## Distribution

*E. vallentini* is endemic to the southern hemisphere with a typically sub-Antarctic circumpolar distribution extending between the Subtropical Convergence and the Antarctic Convergence (John 1936). It has been exceptionally collected also beyond these boundaries (Zimmer 1915; Lomakina 1964 ). Off South America, *E. vallentini* is found in Atlantic and Pacific waters, including southern Chilean fjords.

In Atlantic waters near South America, only spineless individuals have been found (Antezana and Brinton 1981). The geographic distribution of *E. vallentini* in the Eastern South Pacific is divided into 2 areas: an oceanic West Wind Drift area south of 48°S, extending to the Antarctic Convergence, and a coastal area between 38–42°S in the Humboldt Current. *E. vallentini* is concentrated in the area between 53–57°S reaching coastal areas as well, where it is the most abundant sub-Antarctic species (Antezana 1978, 1981). Within the fjords of Chile (including a few stations in the Straits of Magellan), *E. vallentini* was the most frequently encountered and abundant euphausiid accounting for the greatest proportion of the net-zooplankton biomass (Antezana 1976). Guglielmo et al. (1991) and Antezana et al. (1992) also identified it as the dominant euphausiid species within the Straits.

In the present study, both spiny and spineless individuals were observed. Only juveniles (calyptopis and furcilia stages) were sampled in the Pacific area, with total lengths ranging from 5–7 mm (Sts. 5 and 6) to 8–10 mm (Sts. 9 and 11). The number of adults and juveniles progressively increased towards the centre of the Straits, with maximum densities at St. 18. Only a small number of adult specimens were sampled in the Atlantic (St. 26). The species was present at all depths but maximum numbers were recorded from 150–200 m at Central and Pacific stations, where it undergoes ample diel vertical migrations.

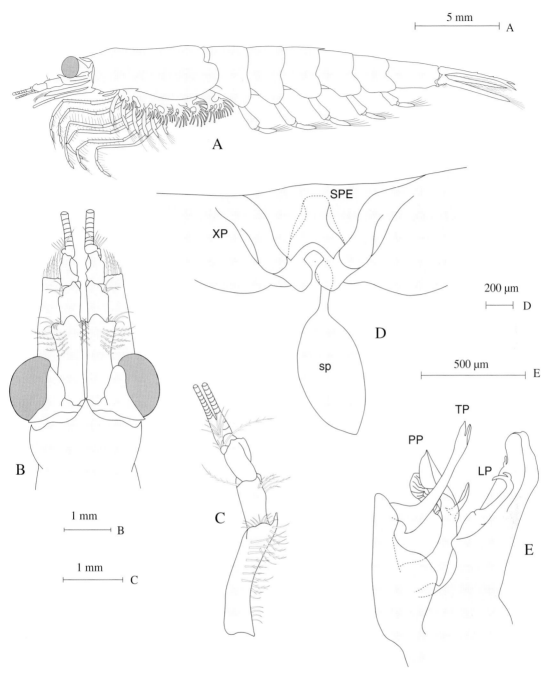

**Fig. 4.1.1A–E.** *Euphausia vallentini.* Female: **A** whole animal, lateral view; **B** front part of carapace and first antennal peduncles; **C** left first antennal peduncle from outer side; **D** copulatory organ. Male: **E** copulatory organ. *LP* Lateral process; *PP* proximal process; *TP* terminal process; *XP* coxal plate; *SPE* sternal plate; *sp* spermatophore

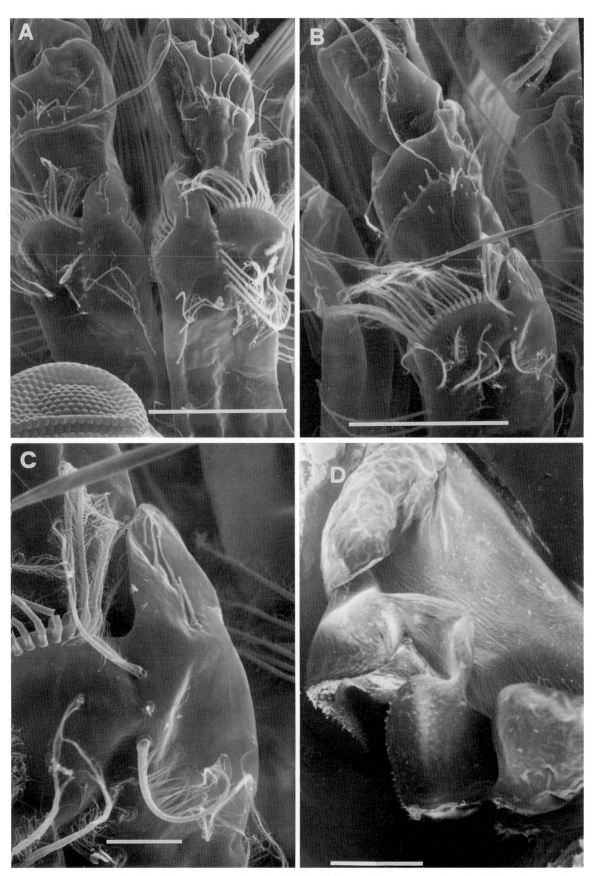

**Fig. 4.1.2A–D.** *Euphausia vallentini.* **A** 1st antennal peduncles dorsal view; **B** left 1st antennal peduncle lateral view; **C** lappet of left 1st antennal peduncle lateral view; **D** thelycum. *Bars* **A,B** 500 μm; **C,D** 100 μm

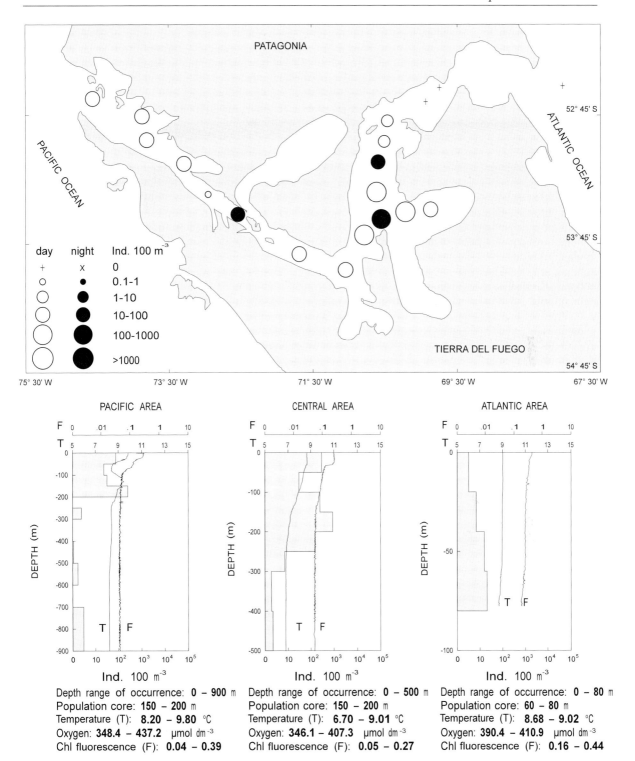

PATAGONIA

PACIFIC OCEAN

ATLANTIC OCEAN

52° 45' S

53° 45' S

54° 45' S

TIERRA DEL FUEGO

| day | night | Ind. 100 m$^{-3}$ |
| + | x | 0 |
| o | • | 0.1-1 |
| O | ● | 1-10 |
| O | ● | 10-100 |
| O | ● | 100-1000 |
| O | ● | >1000 |

75° 30' W     73° 30' W     71° 30' W     69° 30' W     67° 30' W

PACIFIC AREA

Depth range of occurrence: **0 – 900** m
Population core: **150 – 200** m
Temperature (T): **8.20 – 9.80** °C
Oxygen: **348.4 – 437.2** μmol dm$^{-3}$
Chl fluorescence (F): **0.04 – 0.39**

CENTRAL AREA

Depth range of occurrence: **0 – 500** m
Population core: **150 – 200** m
Temperature (T): **6.70 – 9.01** °C
Oxygen: **346.1 – 407.3** μmol dm$^{-3}$
Chl fluorescence (F): **0.05 – 0.27**

ATLANTIC AREA

Depth range of occurrence: **0 – 80** m
Population core: **60 – 80** m
Temperature (T): **8.68 – 9.02** °C
Oxygen: **390.4 – 410.9** μmol dm$^{-3}$
Chl fluorescence (F): **0.16 – 0.44**

**Fig. 4.1.3.** *Euphausia vallentini.* Distribution in the Straits of Magellan in February–March 1991

## 4.2 *Euphausia lucens* Hansen, 1905

*Euphausia lucens* Hansen, 1905b: p. 9 – Hansen, 1911: pp. 26, 27, fig. 8a–d – Tattersall, 1913: p. 876 – Hansen, 1915: p. 84 – Zimmer, 1915: pp. 178, 179 – Colosi, 1917: pp. 183-186, pl. 14 figs. 6–8 – Tattersall, 1924: pp. 19, 20 – Tattersall, 1925: p. 7 – Illig, 1930: p. 499 – Rustad, 1930: pp. 17, 32, 33, fig. 13 a–b – John, 1936: pp. 205–210, figs. 7–11, 30a – Sheard, 1953: tabs. 2-5, 8, 9 – Boden, 1954: p. 208, fig. 11a–c – Boden, 1955: pp. 341, 359–365, figs. 11–16, tabs. 2, 8 – Bary, 1956: pp. 440–442, fig. 8 – Nepgen, 1957: pp. 8, 9, pls. 2, 3 – Mauchline and Fischer, 1969: pp. 64, 65, fig. 21 – Mayer, 1969: pp. 250–255, figs. 2–5 – Ramirez, 1971: pp. 391, 392, pl. 1 fig. 2, pl. 3 fig. 2, pl. 9 – Ramirez, 1973: p. 111 – Antezana et al., 1976: p. 63, figs. 3–12b – Antezana and Brinton, 1981: p. 688, figs. 233–238.
*Euphausia splendens* (partim) Sars, 1885: p. 80, pl. 13, figs. 7-17.
*Euphausia uncinata* Colosi, 1917: p. 186, pl. 14 figs. 9, 10.

Body length –
  females: 11.6–13.2 mm (6 specimens)
  males: 10.4–11.6 mm (4 specimens)

### Description

Lower margin of carapace bearing 1 lateral denticle. Rostrum rather short. Frontal plate short, triangular and as long as wide. Eyes large and spherical. Basal segment of antennular peduncle bearing triangular, pointed to rounded lobe of variable size. Lobe about as long as wide, projected forward and slightly upturned. Third segment bearing robust dorsal keel with straight or slightly curved upper margin along more than 1/2 its length. Antennular scale extending beyond distal margin of eyes. Abdominal segment lacking dorsal spine or processes on distal dorsal margins. Preanal spine bearing 1 or 2 secondary spines, which are larger in females than in males. Subterminal spines of telson bearing denticles on inner margins, giving them a serrated appearance. Terminal process of petasma slender and usually longer than median lobe.

Proximal heel rather developed; distal end bifurcated with slightly expanded inner ramus that is longer and stouter than outer one.

Proximal process shorter and somewhat more robust than terminal process and expanded, on distal end, into 2 membranous leaf-like lappets with fragile edges. These are arranged in different planes and provided with basal septae. Anterior lappet originating closer to base of process and bearing striations. It is shorter and with more fragile edges than the posterior lappet. A secondary spine originates just prior to the implantation of these membranous lappets. This spine is robust, wide at the base and curved downward. Lateral process stout at base and bearing short and robust tooth on its strongly curved distal end. Setiferous lobe bearing 4 plumose seate on distal end. Same lobe also bearing flat, rounded lobe on distal inner side and ear-like lobe near outer side. 2 finger-shaped prominences of the coxal plates are observed laterally. Sternite divided into 2 fairly deep grooves by a median ridge. Opening of pocket delimited at sides of the 2 folds of the coxal plates; these are more rounded in shape anteriorly (Guglielmo and Costanzo, 1978).

## Remarks

*Euphausia lucens* resembles *E. vallentini* (particularly non-spined specimens), due to the rostrum which varies from acute to subacute (Zimmer 1915) and to the variability in the shape and size of the lobe on the 1st antennular segment (Hansen 1911; Zimmer 1915; John 1936; Boden 1955; Bary 1956). As discussed above, the distinction between the 2 species is based on the size and shape of the dorsal keel of the 3rd antennular segment, the denticles on the inner border of the subterminal spines, and the distribution of the setae on the setiferous lobe of the petasma. These 2 species can also be distinguished by the differences in the coxal plates and the sternite, which forms the thelycum (Guglielmo and Costanzo 1978).

## Distribution

*Euphausia lucens* is endemic to the Southern Hemisphere with a typically circumpolar distribution, avoiding the cooler latitudes of the sub-Antarctic (John 1936). Off South America, *E. lucens* is found in Atlantic and Pacific waters, including the fjords of southern Chile. In Atlantic waters near South America, it was collected off the Rio de la Plata estuary to the North of the Antarctic Convergence (Antezana and Brinton 1981). In the Eastern South Pacific, it occupies an area immediately north of *E. vallentini*.

It is concentrated from 41°–45°S and expands its latitudinal range toward the Chilean coast, reaching 37°S to the north and 52°S to the south (Antezana 1978, 1981).

Within the fjords of Chile, it was frequently, but not densely, found at the surface, north of 50°S in waters as saline as 27 ppm (Antezana 1976). It has not been previously reported in the Straits of Magellan (Guglielmo et al. 1991; Antezana et al. 1992). In the present study, it was sampled throughout the Pacific sector up to St. 9, where it was found from 100–200 m; in the Atlantic, it was sampled only at the entrance to the Straits (St. 26) from 0–30 m. Both males and females had a well-developed petasma and thelycum.

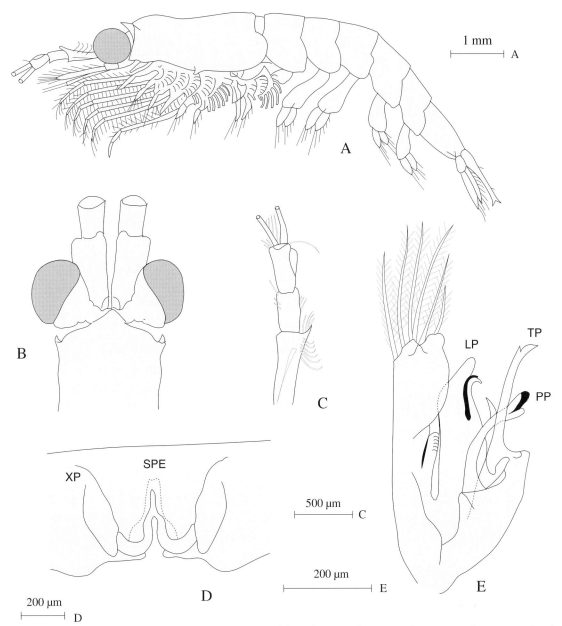

**Fig. 4.2.1A–E.** *Euphausia lucens.* Female: **A** whole animal, lateral view; **B** front part of carapace and 1st antennal peduncles; **C** left 1st antennal peduncle from outer side; **D** copulatory organ. Male: **E** copulatory organ. *LP* Lateral process; *PP* proximal process; *TP* terminal process; *XP* coxal plate; *SPE* sternal plate

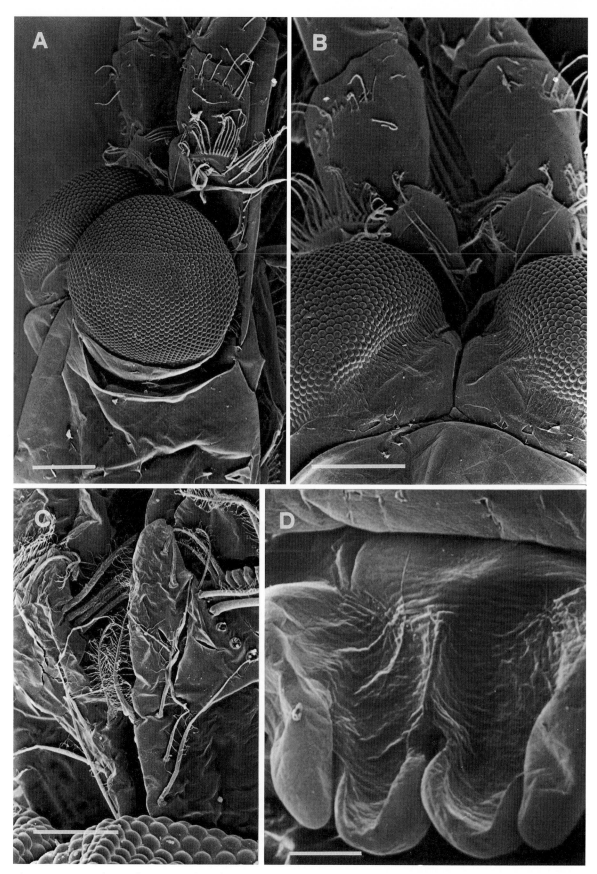

**Fig. 4.2.2A–D.** *Euphausia lucens.* Female: **A** front part of carapace and 1st antennal peduncles lateral view; **B** front part of carapace and 1st antennal peduncles dorsal view; **C** lappet of right 1st antennal peduncle lateral view; **D** thelycum. *Bars* **A,B** 250 μm; **C** 100 μm; **D** 200 μm

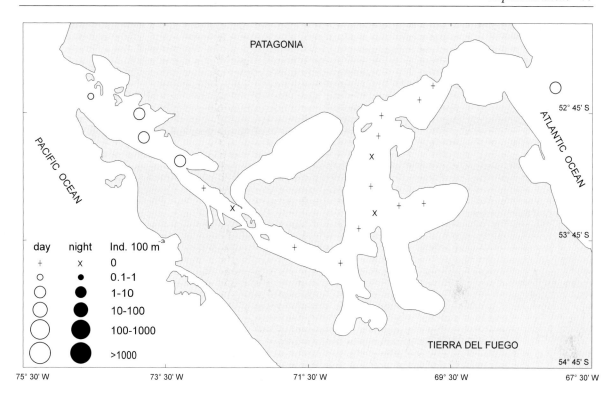

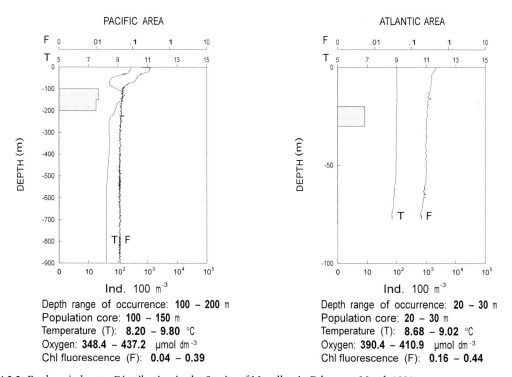

Depth range of occurrence: **100 – 200** m
Population core: **100 – 150** m
Temperature (T): **8.20 – 9.80** °C
Oxygen: **348.4 – 437.2** µmol dm⁻³
Chl fluorescence (F): **0.04 – 0.39**

Depth range of occurrence: **20 – 30** m
Population core: **20 – 30** m
Temperature (T): **8.68 – 9.02** °C
Oxygen: **390.4 – 410.9** µmol dm⁻³
Chl fluorescence (F): **0.16 – 0.44**

**Fig. 4.2.3.** *Euphausia lucens.* Distribution in the Straits of Magellan in February–March 1991

## 4.3 *Thysanoessa gregaria* Sars, 1883

*Thysanoessa gregaria* Sars, 1883: p. 26 – Sars, 1885: p. 120, pl. 21 figs. 8–17, pl. 22 – Hansen, 1905a: p. 25 – Hansen, 1905b: p. 28 – Holt and Tattersall, 1905: p. 115 – Zimmer, 1909: p. 20, figs. 32–34 – Hansen, 1911: pp. 43, 44, fig. 15 – Hansen, 1913: p. 37, pl. 6 fig. 1a–b – Zimmer, 1914: p. 430 – Hansen, 1915: p. 101 – Zimmer, 1915: p. 182 – Tattersall, 1924: pp. 27, 28 – Tattersall, 1925: p. 9 – Tattersall, 1926: p. 22 – Tattersall, 1927: p. 22 – Illig, 1930: p. 526 – Rustad, 1930: pp. 34, 35, fig. 19 – Ruud, 1936: pp. 37-39, fig. 13 – Dakin and Colefax, 1940: pp. 140, 141, fig. 232, pl. 2 fig. 9 – Sheard, 1953: p. 17 et seq., fig. 3, tabs. 2, 6, 8, 9 – Banner, 1954: p. 220, fig. 16a–b – Boden, 1955: p. 344 – Boden et al., 1955: pp. 359–361, fig. 38a–d – Brinton, 1962b: pp. 146-148, figs. 57, 58 – Ponomareva, 1963: p. 26 et seq., fig. 9 – Sheard, 1965: p. 240 – Nemoto, 1966: p. 114 et seq., fig. 2 – Mauchline and Fischer, 1969: pp. 82, 83, fig. 27 – Mauchline, 1971a: p. 7, fig. 9a–d – Mauchline, 1971b: p. 12, fig. 10a–b – Ramírez, 1971: pp. 399, 401, pl. 4 fig. 2, pl. 5 fig. 4, pl. 7 fig. 2, pl. 9 – Talbot, 1974: p. 126 – Antezana et al, 1976: p. 56, figs. 1–4a – Antezana and Brinton, 1981: p. 689, figs. 231-233.

Body length –
    females: 15.2–15.4 mm (2 specimens)
    males: 13.5–13.6 mm (2 specimens)

## Description

Carapace bearing lateral denticle behind midpoint of lower margin; denticle may be missing at times (Nemoto 1966). Rostrum long and rather oblong. Dorsal keel with sharp margin extending over rostrum, frontal plate and gastric region. Eyes very large and transversally strangled in 2 lobes. Upper lobe smaller than lower one. The difference in size is remarkably greater in females. Basal segment of antennular peduncle longer than the sum of the following 2 segments. Flagella are extremely short. Antennal scale reaching about midpoint of 3rd segment of antennular peduncle. 2nd endopodite considerably elongated. Setae from carpus and propodus of 2nd – 4th endopodites as long or longer than those of dactylus. Dorsal keels or epimeral spines present on every abdominal segment; spine on 6th segment much smaller than others. Preanal spine showing sexual dimorphism. In females, it is profusely denticulated and in males it is usually unarmed or exceptionally provided with scarce and minute denticles (Nemoto 1966). Terminal process of petasma broad at base, curved and narrow at middle, and expanded distally to a broad, truncated, rounded and finely serrated edge. Proximal process longer and more slender than terminal process, ending in a short and outwardly curved denticle and in a triangular expansion with serrated distal edge. Spiniform process small, frail and moderately curved. Lateral process somewhat shorter than proximal process, more robust at base and slender and slightly curved at end. Thelycum, only formed by coxal plates, anteriorly resembles a longitudinal section of a funnel and posteriorly forms 2 cavities for the spermatophores (Costanzo and Guglielmo 1976b).

## Remarks

*T. gregaria* closely resembles the deep-water dweller *T. parva*. Clear diagnostic differences between the two species include the petasma and setae from the 1st to the 3rd thoracic endopodites. Identification of the species may be difficult for juvenile and larval stages, which become indistinguishable from those of *T. macrura* and *T. vicina*. The relatively larger gill size has proven to be a good secondary character for *T. gregaria*. Differences between the thelycums of *T. gregaria* and *T. parva* regard only the coxal plates. In *T. gregaria*, they are more extended posterolaterally, delimiting 2 wide pockets, where the spermatophores are located (Costanzo and Guglielmo 1977).

## Distribution

*T. gregaria* inhabits the North (Bigelow 1926; Moore 1952) and South (Zimmer 1914; Ramirez 1971; Antezana and Brinton 1981) Atlantic, the Indian Ocean (Illig 1930) and Mediterranean Sea (Ruud 1936). In both the Pacific and Atlantic Oceans, *T. gregaria* is an antitropical or bipolar species occupying the Transition Zone between Sub polar and Central Water masses in the mid-ocean and between subpolar and equatorial waters in the California and Humboldt Current (Brinton 1962b). Brinton explained the present disjunct distribution of *T. gregaria* resulting from the successive cooling and warming (2.5 °C at 200 m) of the ocean during the ice ages (Brinton 1962b). Off the Chilean coast, *T. gregaria* is found in a wide latitudinal range between 15° and 56°S. In the Humbold Current, it occupies an intermediate oceanic belt 60–200 miles offshore, and in sub-Antarctic waters of the Cape Horn Current, it is concentrated in coastal regions (Antezana 1978, 1981). In Chilean fjords, *T. gregaria* is the second most abundant euphausiid, preceded by *E. vallentini*. Both species have been found in mutually exclusive swarms (Antezana 1976). In the present study, only a small number of specimens were sampled at St. 26. The species seems to be confined to areas at the Atlantic entrance to the Straits, where it was sampled from 20–30 m.

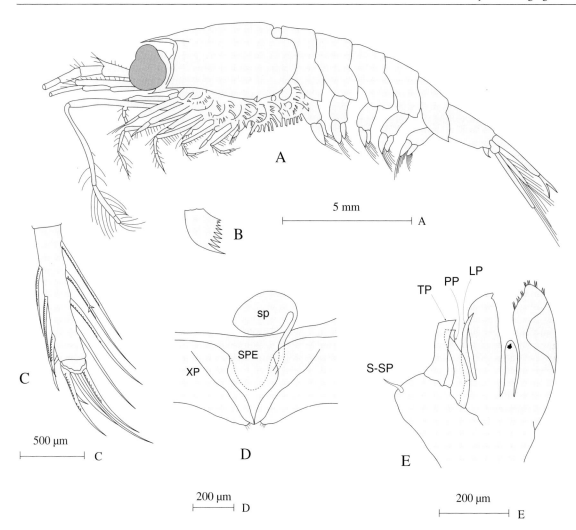

**Fig. 4.3.1A–E.** *Thysanoessa gregaria*. Female: **A** whole animal, lateral view; **B** preanal spine; **C** propodus and dactylus of elongated 2nd thoracic leg; **D** copulatory organ. Male: **E** copulatory organ. *LP* Lateral process; *PP* proximal process; *TP* terminal process; *XP* coxal plate; *SPE* sternal plate; *S-SP* spine-shaped process; *sp* spermatophore

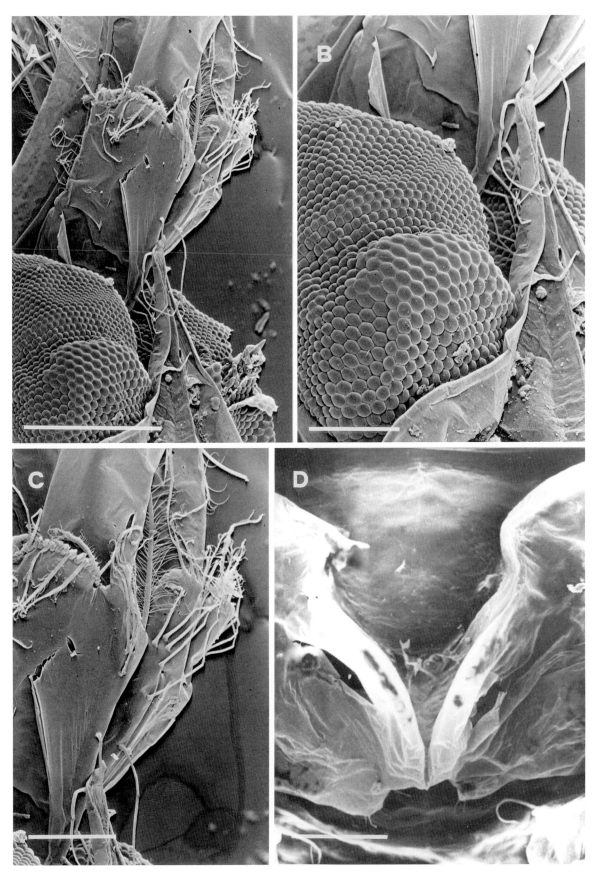

**Fig. 4.3.2A–D.** *Thysanoessa gregaria.* Female: **A** front part of carapace and 1st antennal peduncles lateral view; **B** detail of rostrum lateral view; **C** detail of 1st antennal peduncles lateral view; **D** thelycum. *Bars* **A** 500 µm; **B,C** 200 µm; **D** 100 µm

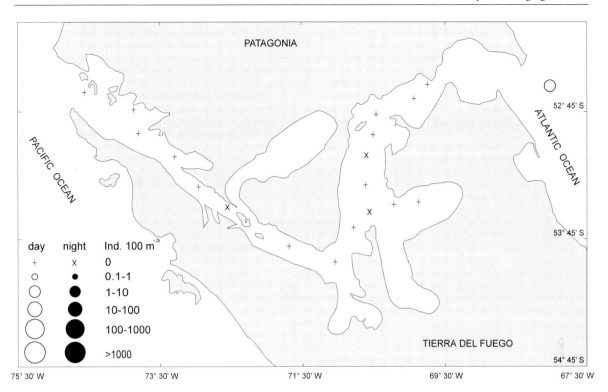

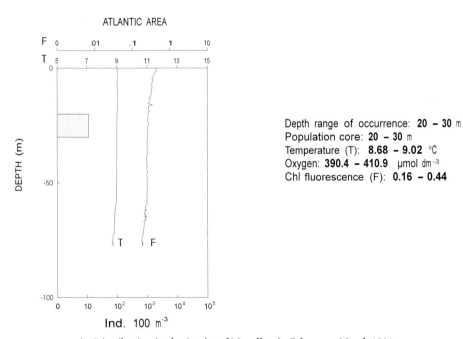

Depth range of occurrence: **20 – 30** m
Population core: **20 – 30** m
Temperature (T): **8.68 – 9.02** °C
Oxygen: **390.4 – 410.9** μmol dm$^{-3}$
Chl fluorescence (F): **0.16 – 0.44**

**Fig. 4.3.3.** *Thysanoessa gregaria.* Distribution in the Straits of Magellan in February–March 1991

## 4.4 *Nematoscelis megalops* Sars, 1883

*N. megalops* Sars, 1883: pp. 27, 28 – Sars, 1885: 127, pl. 23 figs. 5–10, pl. 24 figs. 1-30 – Hansen, 1905a: p. 27 – Holt and Tattersall, 1905: pp. 115, 116 – Hansen, 1908: pp. 90, 91 – Zimmer, 1909: p. 16, figs. 24–27 – Hansen, 1910: p. 106 – Hansen, 1911: pp. 48–50, fig. 18a – Zimmer, 1914: p. 431 – Hansen, 1915: pp. 104–107 – Colosi, 1917: pp. 192, 193 – pl. 16 figs. 43, 44 – Tattersall, 1925: pp. 9, 10 – Tattersall, 1926: p. 22 – Tattersall, 1927: p. 22 – Illig, 1930: pp. 521, 522 – Ruud, 1936: pp. 41, 42, fig. 14 – Einarsson, 1942: pp. 270–272, figs. 5, 6a – Moore, 1952: p. 301, figs. 19, 21 – Sheard, 1953: p. 18 – Boden, 1954: pp. 224–226, fig. 17a-c – Boden, 1955: pp. 343, 370-375, figs. 20–24 – Bary, 1956: pp. 445–447, fig. 13 – Nepgen, 1957: p. 10, pl. 11 – Boden 1961: p. 257 – Brinton, 1962b: pp. 152–154, figs. 62, 63a – Soulier, 1963: pp. 425–428, figs. 11a–e, 12, 13 – Sheard, 1965: p. 240 – Soulier, 1965: p. 183 – Casanova-Soulier, 1968: p. 6 et seq. – Mayer, 1969: pp. 250–252, fig. 1 – Mauchline and Fischer, 1969: pp. 84–86, fig. 28 – Mauchline, 1971a: p. 7, fig. 11a–d – Mauchline, 1971b: p. 12, fig. 11a–k – Ramírez, 1971: pp. 397, 398, pl. 4 fig. 1, pl. 7 fig. 4, pl. 10 – Gopalakrishnan, 1973: pp. 18, 19, pl. 23 – Ramírez, 1973: p. 113 – Gopalakrishnan, 1975: p. 798 et seq., figs. 2–11 – Antezana et al., 1976: p. 55, figs. 1–3a – Antezana and Brinton, 1981: p. 688, figs. 221-233.

Body length –
females: 21.2–23.6 mm (30 specimens)
males: 18.3–20.4 mm (30 specimens)

### Description

Lower margin of carapace lacking lateral denticles. Female rostrum long, slender, curved, acute at the end, and reaching distal margin of eyes. In males, it is short and truncated. Length and shape of rostrum varies widely in males as well as in females. Dorsal keel extending over rostrum, frontal plate and gastric region. Eyes very large, about 1/4 as high as broad, and transversally strangled in two lobes; upper lobe somewhat smaller than lower one. Antennular peduncle shorter and more robust in males than in females. Basal segment expanded on dorsal and distal ends, turned slightly upward without forming a lobe or conspicuous process. Laterodistal outer margin bearing prominent spine. 2nd and 3rd segments lacking spines or processes. Flagella short. Antennal scale narrow and extending beyond midpoint of last antennular segment. Lateral spine of basiscerite poorly developed. Dactylus of 1st endopodite flat and bearing row of ca.11–15 setae on inner margin and 1 on dorsal margin. Propodus bearing row of 6 setae on dorsal margin, 5 setae on mid-dorsal side and 2 on inner margin. 2nd endopodite very elongated and frail, particularly the distal segments. Last 2 segments with bundle of bristles: propodus bearing 2 and dactylus, which is very short, bearing 8. 3rd to 6th endopodites small and with normal number of segments (3 beyond the knee). 7th endopodite bi-segmented in females and absent in males. Eight endopodite rudimentary in both sexes. 4th and 5th abdominal segments each bearing low mid-dorsal keel ending in short spiniform process. Preanal spine usually simple in males and bifurcated in females; this character may be variable. Distal portion of terminal process of petasma longer, thinner and curved in the middle, forming an obtuse angle with proximal portion of process. External edge fringed by row of denticles: proximal ones smaller than distal ones. Proximal process shorter than terminal one and reaching as far as midpoint of serrated margin. Process curved as well, and forming a similar obtuse angle between basal and distal portions. Distal portion rather straight and finely serrated on external edge. Spiniform process slender, straight, stiff and reaching mid-angle of terminal process. Lateral process implanted near base of median lobe sinuous, longer than spine-shaped process, and shorter than proximal process. Setiferous lobe bearing 2 setae on external margin, 3 on internal margin and 2 at tip. Ear-like large lappet extending beyond distal margin of lobe. According to Costanzo and Guglielmo (1976a), proceeding from the posterior margin of the 6th thoracic segment, the sternite extends forward, protruding in a finger-like process and forming a cavity which is anteriorly more accentuated, so that a groove is formed. In this groove arise the stems of the two spermatophores. The coxal plates are highly developed; they bear some setae on the inner surface.

## Remarks

The rostrum varies largely in specimens of both sexes. In females collected off the coast of Chile, it is long and slender, and in males short to absent, but the opposite can also be found exceptionally. The rostrum is rarely reduced in males from Atlantic South African (Boden 1954) and Mediterranean (Boden 1954; Soulier 1963) waters. Gopalakrishnan (1975) remarked that the rostrum of males of the genus *Nematoscelis* is only rarely variable.

The side of the carapace has a sinuous and delicate cervical groove overlapping the bucal one. This latter groove is also sinuous but deeper than the cervical one. This is best observed from a dorsal view. This feature could be an additional character to distinguish *N. megalops* from the very similar species *N. difficilis* Hansen. Another character of similar importance is the presence of low mid-dorsal keels ending in posterior spines along the 4th and 5th abdominal segments. This has also been observed in specimens off New Zealand (Bary 1956). The preanal spine, bifurcated in females and simple in males, can vary to a great extent. *N. megalops* may be separated from *N. difficilis* (Hansen 1911) by the structure of the petasma. Females and subadults are difficult to differentiate. In fact, Einarsson

(1942), Karedin (1971) and Costanzo an Guglielmo (1980) considered both as identical species based on their petasma and thelycum. However, Boden (1954), Boden et al. (1955), Bary (1956), Mayer (1969) and Gopalakrishnan (1975) have all recognized the validity of both species.

## Distribution

*N. megalops* inhabits tropical and subtropical regions of the Indian Ocean (Illig 1930) and Mediterranean Sea (Ruud 1936; Soulier 1963). In the North Atlantic, it is found in sub-Antarctic and subtropical waters (Moore 1952). In the South Atlantic, it occurs between sub-Antarctic and subtropical waters of the eastern sector (Boden 1955; Nepgen 1957) and between 45° and 49° S off the coast of Argentina (Ramirez 1971; Antezana and Brinton 1981). In the South Pacific, *N. megalops* is found in the Transition Zone, a transoceanic belt between equatorial and sub-Antarctic waters, extending from the South American coast to New Zealand and Australia (Brinton 1962b). In the Eastern South Pacific off Chile, *N. megalops* is found in the oceanic waters of the Humboldt Current and in the most coastal sub-Antarctic waters of the Cape Horn Current. The geographical distribution and overall latitudinal range of *N. megalops* is similar to the other Transition Zone species *T. gregaria*, but it is somewhat more restricted, encompassing a still wider (24°–55°S) latitudinal range (Antezana 1978, 1981). *N. megalops* has also been found along the entire Chilean fjord region, with the exception of a few locations (e.g. those of the Straits of Magellan; Antezana 1976). Brinton (1962b) suggested that the tip of South America is a land barrier preventing the passage of most antitropical species, with some exceptions such as *T. gregaria*. The above mentioned records provide evidence of the connection between Pacific and Atlantic populations.

In the present study, *N. megalops* was the second most abundant species sampled in the Straits after *E. vallentini*. It was present at almost all stations even though the number of adults sampled was small. Most specimens were juvenile or furcilia stages (4–6 mm). In the Pacific and Central areas, where it was more common, the species was sampled from the surface to 300 m.

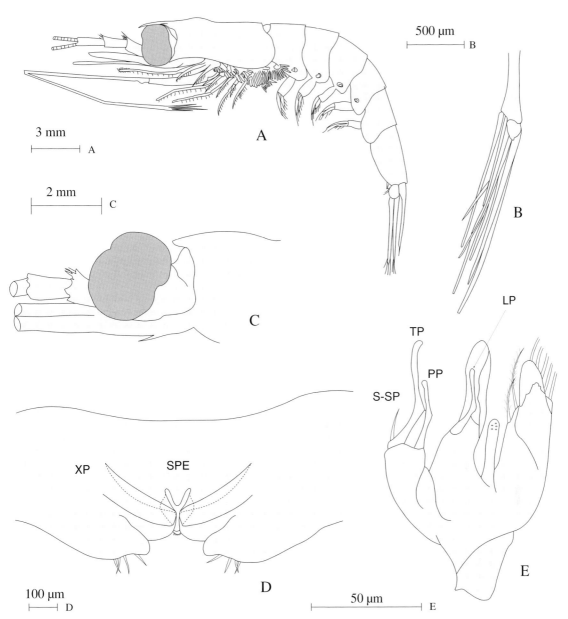

**Fig. 4.4.1A–E.** *Nematoscelis megalops.* Female: **A** whole animal, lateral view; **B** end of propodus and dactylus of elongated 2nd thoracic leg; **C** copulatory organ. Male: **D** head region; **E** copulatory organ. *LP* Lateral process; *PP* proximal process; *TP* terminal process; *XP* coxal plate; *SPE* sternal plate; *S-SP* spine-shaped process

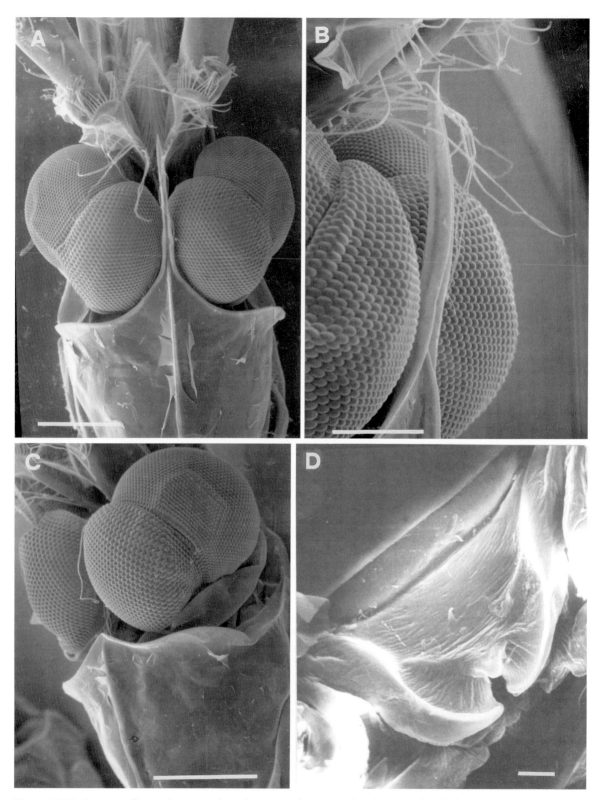

**Fig. 4.4.2A-D.** *Nematoscelis megalops*. Female: **A** front part of carapace dorsal view; **B** detail of rostrum lateral view. Male: **C** front part of carapace dorsal-lateral view. Female: **D** thelycum. *Bars* **A,C** 1 mm; **B** 500 µm; **C,D** 100 µm

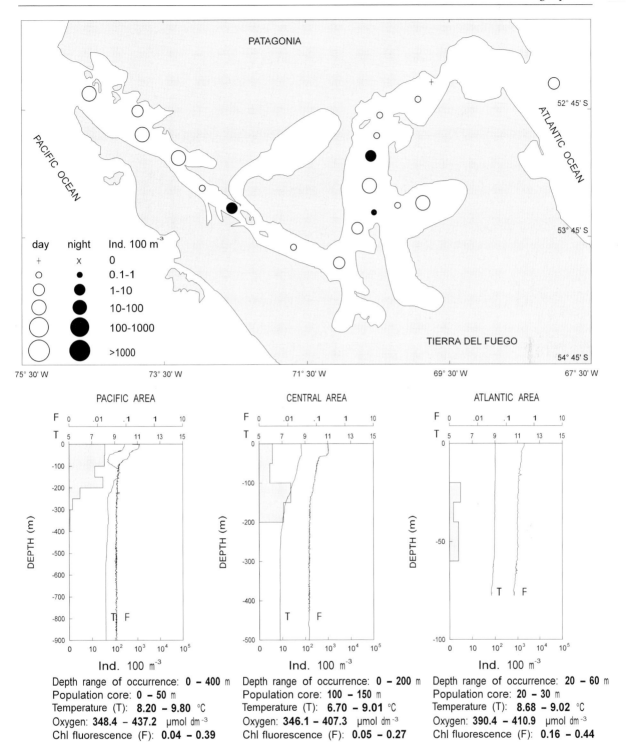

Depth range of occurrence: **0 – 400** m
Population core: **0 – 50** m
Temperature (T): **8.20 – 9.80** °C
Oxygen: **348.4 – 437.2** µmol dm⁻³
Chl fluorescence (F): **0.04 – 0.39**

Depth range of occurrence: **0 – 200** m
Population core: **100 – 150** m
Temperature (T): **6.70 – 9.01** °C
Oxygen: **346.1 – 407.3** µmol dm⁻³
Chl fluorescence (F): **0.05 – 0.27**

Depth range of occurrence: **20 – 60** m
Population core: **20 – 30** m
Temperature (T): **8.68 – 9.02** °C
Oxygen: **390.4 – 410.9** µmol dm⁻³
Chl fluorescence (F): **0.16 – 0.44**

**Fig. 4.4.3.** *Nematoscelis megalops.* Distribution in the Straits of Magellan in February–March 1991

## 4.5 *Stylocheiron longicorne* Sars, 1883

*S. longicorne* Sars, 1883: p. 32 – Sars, 1885: pp. 144, 145, pl. 27 fig. 5 – Hansen, 1908: p. 93 – Hansen, 1910: pp. 120, 121, pl. 16 fig. 5a–b – Hansen, 1912: pp. 279, 280, pl. 11 fig. 4a–b – Zimmer, 1914: pp. 436, 437 – Hansen, 1916: p. 652 – Colosi, 1917: pp. 197, 198, pl. 15 figs. 39–42 – Tattersall, 1924: p. 30 – Tattersall, 1927: pp. 26, 27 – Illig, 1930: p. 527 – Ruud, 1936: p. 15 et seq., figs. 5, 17 – Tattersall, 1939: pp. 222, 223 – Banner, 1950: pp. 37, 38, pl. 4 fig. 5 – Boden, 1954: pp. 234–236, fig. 21a–c – Boden et al., 1955: p. 388, fig. 53a–c – Boden, 1955: p. 345 – Lewis, 1955: p. 197, fig. 5 – Nepgen, 1957: p. 11, pl. 12 – Boden, 1961: p. 260 – Roger, 1968: p. 38, fig. 15, tab.1 – Mauchline and Fischer, 1969: pp. 98, 100, fig. 35 – Mauchline, 1971a: p. 7, fig.14a – Mauchline, 1971b: p. 13, fig. 14 a–g – Ramirez, 1971: pp. 402–404, pl. 1 fig. 4, pl. 4, fig. 3 – Brinton, 1975: pp. 215, 216, figs. 115, 124a–f – Antezana and Brinton, 1981: pp. 689, 694, figs. 228–233.
*S. mastigophorum* Chun,1887: p. 30 (partim).

Body length –
   female : 7.6 mm (1 specimen)
   male: 7.2 mm (1 specimen)

## Description

Frontal plate of carapace triangular and projected in very thin and short rostrum. Gastric area bearing short and abrupt mid-dorsal keel. Eyes large and constricted into two lobes that are twice as high as wide. Upper lobe as wide or somewhat wider than lower lobe and with numerous crystal, elongated cones. Antennular peduncle in males as long as carapace. Basal segment similar in both sexes, lacking lobes or similar processes. Distal segments differ between sexes, being much longer and more slender in females than in males. 3rd endopodite greatly elongated, ending in false chela with numerous spines on propodus and dactylus. Abdominal segments lacking spines or keels. 6th abdominal segment longer than high; length and height varies considerably. Terminal process of petasma somewhat broader at base, with distal half forming obtuse angle with proximal half. Distal edge bearing 6 tiny denticles. Proximal process as long and thick as terminal process and with smooth edge. Spiniform process robust and somewhat curved. Lateral process smaller and less curved than proximal and terminal processes. Thelycum consists entirely of the coxal plates forming a pocket which is delimited ventrally by a triangular opening. Two spermatophores are attached to the lateral surface of the pocket bottom. 4 internal and 5 external bristles are inserted on the front edge of the coxal plates (Costanzo and Guglielmo 1976a).

## Remarks

The ratio of the width of the upper lobe/lower lobe and that of the height/length of the 6th abdominal segment vary considerably within the species. Brinton (1962a) used these characters, along with the number of crystal cones, to identify long and short forms of *S. longicorne*. Specimens from colder waters off Chile resemble long forms even though they have 7–8 crystal cones, which corresponds to the number of cones in the short form.

## Distribution

*S. longicorne* is an equatorial, subtropical epipelagic species (Brinton 1962b). It is found in the Atlantic and Pacific Oceans, as well as in the

Mediterranean Sea. In the Pacific, it is widely distributed between 45°N and 45°S. Both short and long forms are found in the Southeastern Pacific. Off Chile, *S. longicorne* has been found along the coast between 25° and 38°S, and occasionally between 20° and 48°S (Antezana 1978). The Southern most record of the species is off the Patagonian shelf at 48°S and 91°W. It was not found before in the Chilean nothern fjords nor along their entire latitudinal range (Antezana et al. 1992). In the South Atlantic off Argentina, *S. longicorne* was found toward the subtropical sectors (Antezana and Brinton 1981). In the present study, few specimens were sampled including one female from 50–100 m at St. 9 and two males from 100–150 m at St. 14. They all resembled the "long form" described by Brinton (1962a).

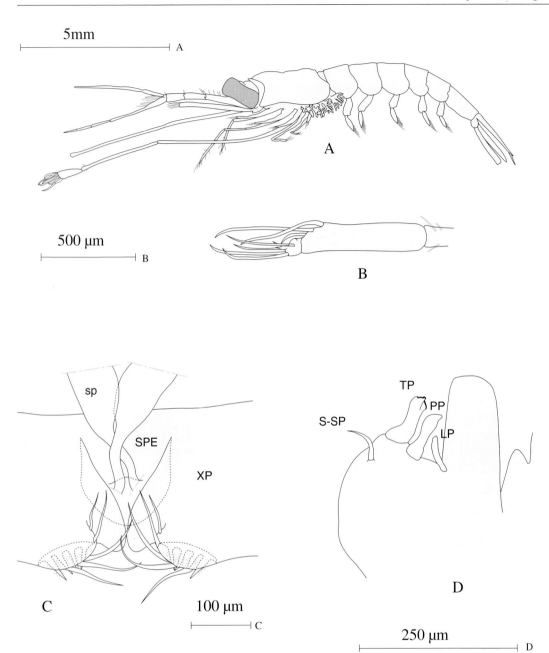

**Fig. 4.5.1A–D.** *Stylocheiron longicorne*. Female: **A** whole animal, lateral view; **B** false chela of elongated 3rd thoracic leg; **C** copulatory organ. Male: **D** copulatory organ. *LP* Lateral process; *PP* proximal process; *TP* terminal process; *XP* coxal plate; *SPE* sternal plate; *S-SP* spine-shaped process; *sp* spermatophore

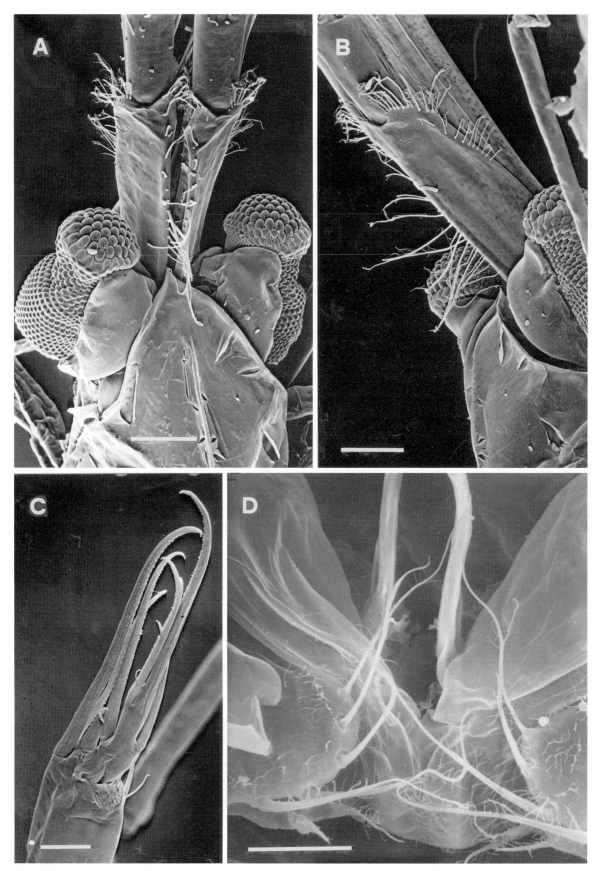

**Fig. 4.5.2A–D.** *Stylocheiron longicorne.* Female: **A** front part of carapace and 1st antennal peduncles dorsal view; **B** front part of carapace and right 1st antennal peduncle lateral view; **C** false chela of elongated 3rd thoracic leg; **D** thelycum. *Bars* **A,B** 250 μm; **C,D** 100 μm

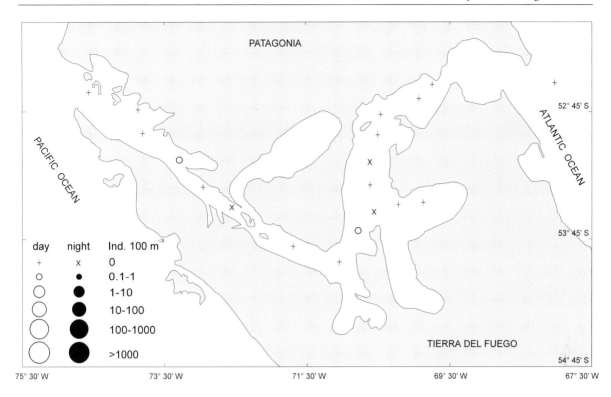

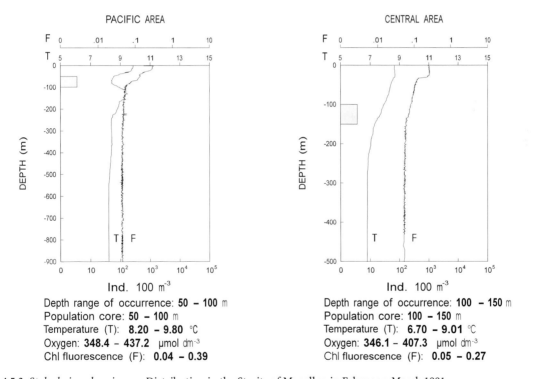

PACIFIC AREA

Depth range of occurrence: **50 – 100** m
Population core: **50 – 100** m
Temperature (T): **8.20 – 9.80** °C
Oxygen: **348.4 – 437.2** μmol dm⁻³
Chl fluorescence (F): **0.04 – 0.39**

CENTRAL AREA

Depth range of occurrence: **100 – 150** m
Population core: **100 – 150** m
Temperature (T): **6.70 – 9.01** °C
Oxygen: **346.1 – 407.3** μmol dm⁻³
Chl fluorescence (F): **0.05 – 0.27**

**Fig. 4.5.3.** *Stylocheiron longicorne*. Distribution in the Straits of Magellan in February–March 1991

## 4.6 *Stylocheiron maximum* Hansen, 1908

*S. maximum* Hansen, 1908: p. 92 – Hansen, 1910: pp. 121, 122, pl. 16 fig. 6a–d – Hansen, 1912: p. 283 – Zimmer, 1914: p. 438 – Hansen, 1915: p. 12 – Hansen, 1916: p. 654 – Tattersall, 1925: p. 11 – Tattersall, 1926: p. 3 – Tattersall, 1927: p. 27 – Illig, 1930: p. 53 – Ruud, 1936: p. 16 et seq., figs. 5–18 – Banner, 1950: pp. 39–42, pl. 4 fig. 26a–j – Banner, 1954: p. 42 – Boden, 1954: p. 237, fig. 22a–c – Boden, 1955: p. 344 – Boden et al., 1955: pp. 391–393, fig.55 – Nepgen, 1957: p. 11, pl. 12 – Brinton, 1962a: p. 169, figs. 81, 82 – Brinton, 1962b: p. 176 – Casanova-Soulier, 1968: et seq. – Roger, 1968: p. 38, fig. 13, tab. 1 – Mauchline and Fischer, 1969: p. 102, fig. 36 – Mauchline, 1971a: p. 7, figs. 13b, 14b, 15b – Mauchline, 1971b: p. 13, fig. 13a–f – Brinton, 1975: p. 218, figs. 116, 124d – Antezana et al., 1976: p. 55, figs. 1–2a – Antezana and Brinton, 1981: p. 689, figs. 229–233.

Body length –
   maximum length of females: 17.8 mm

### Description

Frontal plate of carapace projected on long and pointed rostrum, extending beyond anterior margin of eyes. Gastric region, delimited by conspicuous cervical groove, bulged and supporting a low, short and abrupt dorsal keel. Eyes large and transversally constricted at middle into upper and lower lobes; lower lobe varies from being equal, narrower or broader than the other. Eyes lacking oblong crystal cones. Basal segment of antennular peduncle similar in males and females. 2nd segment as short and robust as 3rd in males. Both are elongated and slender in females; 3rd is almost twice as long as 2nd. Antennular flagella of males cylindrical or somewhat flattened, with globose basal segment bearing several long setae. 3rd endopodite greatly elongated, ending in a false claw formed by a finger-like curved process on distal margin of propodus and dactylus; this in turn projects into long curved spine, and another 2–3 secondary spines. Sixth abdominal segment 2.2.–2.4 times longer than high. Terminal process of petasma somewhat longer and more robust than proximal process. Both are sinuous. Spiniform process long and slender. Lateral process straight and very short.

### Remarks

According to Hansen (1910), *S. maximum* and *S. abbreviatum* form a natural group of similar species. Brinton (1962b) also included *S. robustum* to this group, based on several common features. In fact, these 3 species are quite difficult to distinguish, particularly sub-adult specimens. The following features, suggested by Brinton (1962a), may help in their diagnosis:

1. The upper lobe of the eyes is narrower in *S. robustum* than in *S. maximum,* and is even narrower in *S. abbreviatum.*
2. The length/height relation of the 6th abdominal segment is 2.4 in *S. maximum,* 1.65–2.0 in *S. abbreviatum* and 1.7-1.8 in *S. robustum.*
3. The terminal process ends in an outward curved process and is shorter than the proximal process. This is not the case in *S. maximum* and *S. abbreviatum.*
4. The lateral process is very short in *S. maximum.*
5. The largest size and largest eyes characterize *S. maximum,* while *S. robustum* is smaller and stockier, and has smaller eyes.

According to Costanzo and Guglielmo (1976b), in these 2 species of the maximum-group, the thelycum is likewise formed by the coxal plates only. The large pocket is more widely opened in *S. maximum*, where the edges of the coxal plates are connecting near the bottom of the pocket, whereas in *S. abbreviatum* they are connecting earlier. In both species, several bristles are implanted along the frontal margin of the coxal plates.

## Distribution

*S. maximum* is a mesopelagic cosmopolitan species except that it does not occur in polar regions. It is widely found in the Atlantic and Indian Oceans (Hansen 1910, 1912; Zimmer 1914; Il-lig 1930; Tattersall 1939; Ponomareva 1963) and Mediterranean Sea (Ruud 1936; Casanova-Soulier 1968). In the Pacific, *S. maximum* occupies a wide range between 60°N (Gulf of Alaska) to 63°S in the western South Pacific. Around South America, it has been found off Argentina (Antezana and Brinton 1981). Off Chile, solitary subadult and adult specimens have been randomly sampled throughout the Humboldt and Cape Horn Currents, as far as 60°S (Antezana 1978) and along the southern fjords (Antezana 1976).

Although considered a mesopelagic species, larval stages, juveniles and occasionally adults of *S. maximum* have been found in the upper 200 m (Brinton 1962b; Antezana 1976, 1978). In the present study, only one female was sampled from 200–300 m at St. 15.

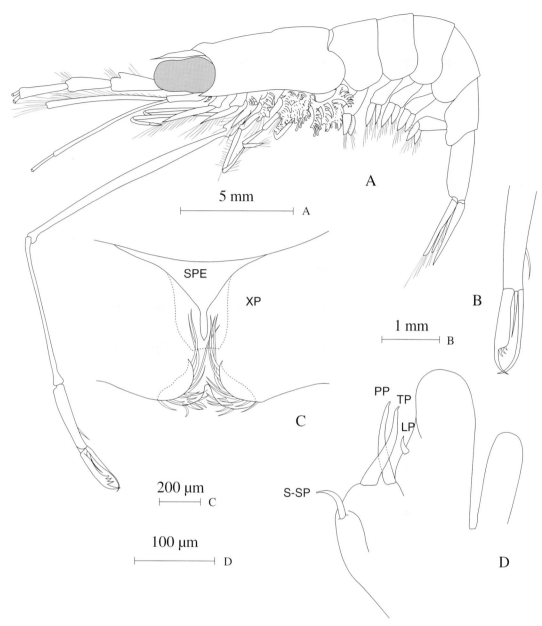

**Fig. 4.6.1A–D.** *Stylocheiron maximum.* Female: **A** whole animal, lateral view; **B** true chela of elongated 3rd thoracic leg; **C** copulatory organ. Male: **D** copulatory organ. *LP* Lateral process; *PP* proximal process; *TP* terminal process; *XP* coxal plate; *SPE* sternal plate; *S-SP* spine-shaped process

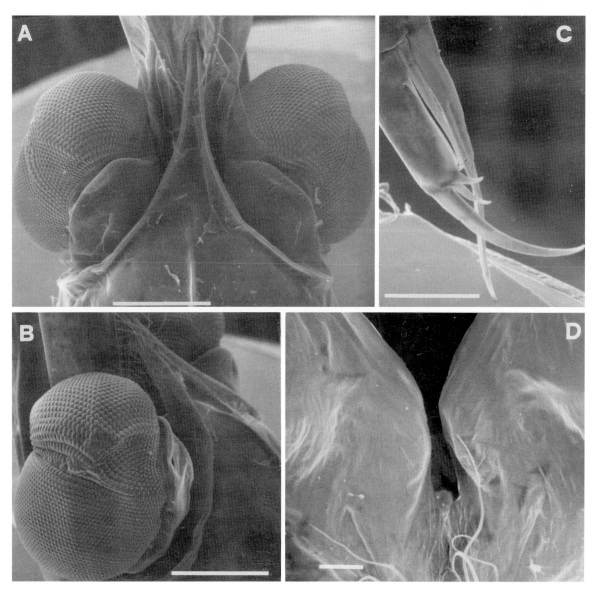

**Fig. 4.6.2A–D.** *Stylocheiron maximum.* Female: **A** front part of carapace dorsal view; **B** detail of head lateral view; **C** true chela of elongated 3rd thoracic leg; **D** thelycum. *Bars* **A,B** 1 mm; **C** 500 μm; **D** 100 μm

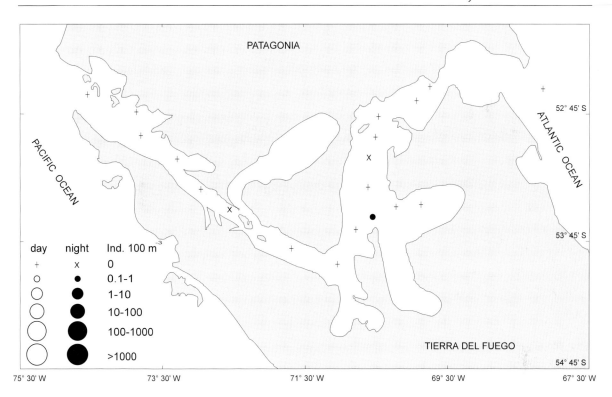

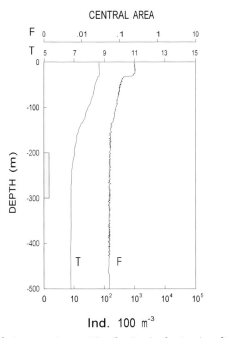

Depth range of occurrence: **200 – 300** m
Population core: **200 – 300** m
Temperature (T): **6.70 – 9.01** °C
Oxygen: **346.1 – 407.3** μmol dm⁻³
Chl fluorescence (F): **0.05 – 0.27**

**Fig. 4.6.3.** *Stylocheiron maximum.* Distribution in the Straits of Magellan in February–March 1991

# References

Antezana T (1976) Diversidad y equilibrio ecológico en comunidades pelágicas. In: Orrego F (ed) Preservación del medio ambiente marino. Inst Est Internac U de Chile, Santiago, pp 40–54

Antezana T (1978) Distribution of euphausiids in the Chile-Peru Current with particular reference to the endemic *E. mucronata* and the oxygen minimun layer. PhD Thesis, Scripps Institution of Oceanography, University of California, San Diego 476 pp

Antezana T (1981) Zoogeography of euphausiids of the South Eastern Pacific Ocean. Memorias del seminario sobre indicadores biológicos del plancton. UNESCO, Montevideo, pp 5–23

Antezana T (1985) Euphausiids. In: Fisher W, Hureau JC (eds) FAO Species identification sheets for fishery purposes. Southern Ocean, FAO, Rome 1:71–871

Antezana T, Brinton E (1981) Euphausiacea. In: Boltowskoy D (ed) Atlas del zooplacton del Atlántico Sudoccidental. INIDEP, Buenos Aires, pp 681–698

Antezana T, Aguirre N, Bustamante R (1976) Clave ilustrada y distribución latitudinal de las especies de krill del Océano Antártico (Crustacea, Zooplancton). Ser Cient Inst Antart Chil 4: 53–69

Antezana T, Guglielmo L, Ghirardelli E (1992) Microbasins within the Strait of Magellan affecting zooplankton distribution. In: Gallardo VA, Ferretti O, Moyano HI (eds) Oceanografía en Antártica. ENEA Proyecto Antárctica, Concepcion, Chile, pp 459–466

Banner AH (1950) A taxonomic study of the Mysidacea and Euphausiacea (Crustacea) of the northeastern Pacific. Part III. Euphausiacea. Trans R Can Inst 28: 1–62

Banner AH (1954) New records of Mysidacea and Euphausiacea from the northeastern Pacific and adjacent areas. Pac Sci 8: 125–139

Bary BM (1956) Notes on ecology, systematics and development of some Mysidacea and Euphausiacea (Crustacea) from New Zealand. Pac Sci 10: 431– 467

Bigelow HB (1926) Plankton of the offshore waters of the Gulf of Maine. Bull Bur Fish Wash 40: 1–509

Boden BP (1954) The euphausiid crustaceans of southern African waters. Trans R Soc S Afr 34: 181–243

Boden BP (1955) Euphausiacea of the Benguela Current. Discovery Rep 27: 337–376

Boden BP (1961) Euphausiacea (Crustacea) from tropical West Africa. Atl Rep 6: 251–262

Boden BP, Johnson MW, Brinton E (1955) The Euphausiacea (Crustacea) of the North Pacific. Bull Scripps Inst Oceanogr 6: 287–400

Brinton E (1962a) Two new species of Euphausiacea, *Euphausia nana* and *Stylocheiron robustum,* from the Pacific. Crustaceana 4: 167–179

Brinton E (1962b) The distribution of Pacific euphausiids. Bull Scripps Inst Oceanogr 8: 51–270

Brinton E (1975) Euphausiids of Southeast Asian waters. Naga Expedition. Naga Rep 4: 1–287

Casanova-Soulier B (1968) Les euphausiacées de la Mediterranée. Comm Int Explor Scient Mer Mediterr, Comite du Plancton, Monaco, pp 62

Chun C (1887) Die pelagische Thierwelt in grösseren Meerestiefen und ihre Beziehung zu der Oberflächenfauna. Zoologica (Stuttg) 1: 1–66

Colosi G (1917) Crostacei. Part II. Eufausiacei. Raccolte planctoniche fatte dalla R. Nave Liguria 2: 165–205

Costanzo G, Guglielmo L (1976a) Diagnostic value of the thelycum in euphausiids. I. Mediterranean species (first note). Crustaceana 31: 45–53

Costanzo G, Guglielmo L (1976b) Diagnostic value of the thelycum in euphausiids. I. Mediterranean species (second note). Crustaceana 31: 178–180

Costanzo G, Guglielmo L (1977) Sur l'importance du thelycum dans la systematique des espèces voisines du genre *Thysanoessa* (Euphausiacea). Rapp Comm Int Mer Médit 24: 129–130

Costanzo G, Guglielmo L (1980) Diagnostic value of the thelycum in euphausiids. I. Oceanic species. Genus *Nematoscelis*. Mar Biol 56: 311–317

Dakin WJ, Colefax AN (1940) The plankton of the Australian coastal waters of New South Wales 1. Publ Univ Sydney Sept Zool Monogr 1: 1–215

Einarsson H (1942) Notes on Euphausiacea I-II. On the systematic value of the spermatheca, on sexual dimorphism in *Nematoscelis* and on the male in *Bentheuphausia*. Vidensk Medd Dans Naturhist Foren 106: 263–286

Gopalakrishnan K (1973) Development and growth studies of the euphausiid *Nematoscelis difficilis* (Crustacea) based on rearing. Bull Scripps Inst Oceanogr 20:1–87

Gopalakrishnan K (1975) Biology and taxonomy of the genus *Nematoscelis* (Crustacea, Euphausiacea). Fish Bull US 73: 797-814

Guglielmo L, Costanzo G (1978) Diagnostic value of the thelycum in euphausiids. II. Oceanic species. Genus *Euphausia* Dana, 1852. Arch Oceanogr Limnol 19: 143–155

Guglielmo L, Antezana T, Costanzo G, Zagami G (1991) Zooplankton communities in the Straits of Magellan. Mem Biol Mar Oceanogr 19:157–161

Hansen HJ (1905a) Preliminary report on the Schizopoda collected by H.S.H. Prince Albert of Monaco during the cruise of the *Princess Alice* in the year 1904. Bull Mus Oceanogr Monaco 30: 1–32

Hansen HJ (1905b) Further notes on the Schizopoda. Bull Mus Oceanogr Monaco 42: 1–32

Hansen HJ (1908) Crustacea Malacostraca. Dan Ingolf Exped 3: 1–120

Hansen HJ (1910) The Schizopoda of the Siboga Expedition. Siboga Exped 37: 1–123

Hansen HJ (1911) The genera and species of the order Euphausiacea, with account of remarkable variation. Bull Inst Océanogr Monaco 210: 1–54

Hansen HJ (1912) Reports on the scientific results of the expedition to the eastern tropical Pacific, in charge of Alexander Agassiz, by the U.S. Fish Commission steamer *Albatross*, from October 1904 to March 1905. Lieut. Commander L.M. Garrett, U.S.N. commanding. 27. Schizopoda Mem Mus Comp Zool Harv 35: 175–296

Hansen HJ (1913) Crustacea Schizopoda. Rep Swed Antarct Exped 1901–1903 1:56

Hansen HJ (1915) The Crustacea Euphausiacea of the United States Museum. Proc US Natl Mus 48: 59–114

Hansen HJ (1916) The Euphausiacean crustaceans of the *Albatross* expedition to the Philippines. Proc US Natl Mus 49: 635,654

Holt EWL, Tattersall WM (1905) Report on the schizopods collected by Mr. George Murray, F.R.S., during the cruise of the *Oceana* in 1889. Ann Mag Nat Hist Ser 7: 1–10

Holt EWL, Tattersall WM (1906) Preliminary notice of the Schizopoda collected by H.M.S. *Discovery* in the Antarctic region. Ann Mag Nat Hist Ser 7: 1–11

Illig G (1930) Die Schizopoden der Deutschen Tiefsee -Expedition. Rep Valdivia Exped 22: 397–625

John DD (1936) The southern species of the genus *Euphausia*. Discovery Rep 14: 193–324

Karedin E P (1971) On the similarity between *Nematoscelis megalops* G.O. Sars 1885 and *N. difficilis* Hansen 1911 (euphausiacean Crustacea) and on the substraction from the isolation of *N. difficilis* Hansen into a separate species. Pac Sci Res Inst Fish Oceanogr (TINRO) 75: 121–129

Lewis JB (1955) Some larval euphausiids of the genus *Stylocheiron* from the Florida Current. Bull Mar Sci Gulf Caribb 5: 190–202

Lomakina NB (1964) Euphausiids fauna of the Antarctic and notalian regions Res. Biol iss Sov Antarkt Eksp 2: 254–334 (in Russian)

Mauchline J (1971a) Euphausiacea adults. Cons Int Explor Mer Zooplankton sheet 134: 1–8

Mauchline J (1971b) Euphausiacea larvae. Cons Int Explor Mer Zooplankton sheet 135/137: 1–16

Mauchline J, Fischer LR (1969) The biology of Euphausiids. Adv Mar Biol 7, New York, Academic Press, 454 pp.

Mayer R (1969) Quelques donnes sur les euphausiidacés (crustaces) de l'Atlantique-sud et du Pacifique du sudest. Ann Mus Civ Stor Nat Giacomo Doria 77: 250–256

Moore HB (1952) Physical factors affecting the distribution of euphausiids in the North Atlantic. Bull Mar Sci Gulf Caribb 1: 278–305

Nemoto T (1966) *Thysanoessa* euphausiids, comparative morphology, allomorphosis and ecology. Sci Rep Whales Res Inst Tokyo 20: 109–155

Nepgen CS de V (1957) The euphausiids of the west coast of S. Africa. Invest Rep Div Fish Un S Afr 28: 1–30

Ponomareva LA (1963) The euphausiids of the North Pacific, their distribution and ecology. Dokl Akad Nauk SSR, 1–142. Israel Progr for Sci Transl 1966.

Ramírez F (1971) Eufáusidos de algunos sectores del Atlántico Sud-occidental. Physis 30: 385–405

Ramírez F (1973) Eufáusidos de la campaña oceanográfica *Walter Herwig*. 1966. Physis 32: 105–114

Roger C (1968) Euphausiacés des Tuamotus. Cah ORSTOM Ser Oceanogr 6: 31–38

Rustad D (1930) Euphausiacea with notes on their biogeography and development. Sci Results Norw Antarct Exped 5: 1–82

Ruud J T (1936) Euphausiacea. Rep Dan Oceanogr Exped Mediterr (Biol) 2: 1–86

Sars GO (1883) Prelimany notices on the Schizopoda of H.M.S. *Challenger* Expedition Forh Vidensk Selks Krist 7: 1–43

Sars GO (1885) Report on the Schizopoda collected by H.M.S. *Challenger* Expedition during the years 1873-1876. The Voyage of H.M.S. *Challenger* Report, Zoology 13: 1–228

Sheard K (1953) Taxonomy, distribution and development of the Euphausiacea (Crustacea). Rep BANZ Antarct Res Exped Ser B (Zool Bot) 8: 1–72

Sheard K (1965) Species group in the zooplankton of eastern Australian slope waters 1938-1941. Austr J Mar Freshw Res 16: 219–254

Soulier B (1963) Pêches planctoniques, superficielles et profondes, en Méditerranée occidentale. IV. Euphausiacés. Rev Trav Inst (Sci Tech) Peches Marit 27: 417–440

Soulier B (1965) Euphausiacés des bancs de Terre-Neuve de Nouvelle-Ecosse et du Golfe du Maine. Rev Trav Inst (Sci Tech) Peches Marit 29: 173–190

Stebbing TRR (1900) On some crustaceans from the Falkland Islands collected by Mr. Rupert Vallentin. Proc Zool Soc Lond: 517–568

Talbot MS (1974) Distribution of euphausiid crustaceans from the Agulhas Current. Zool Afr 9: 93–145

Tattersall WM (1908) Crustacea VII. Schizopoda. Nat Antarct Exped 1901-1904, Nat Hist (Zool) 4: 1–42

Tattersall WM (1913) The Schizopoda, Stomatopoda and non Antarctic Isopoda of the Scottish National Antarctic Expedition. Trans R Soc Edinb 49: 865–894

Tattersall WM (1924) Crustacea. VIII Euphausiacea. Br Antarct Terra Nova Exped 1910, Zool 8: 1–36

Tattersall WM (1925) Mysidacea and Euphausiacea of marine survery, South Africa. Rep Fish Mar Biol Sur Un S Afr No. 4 (1924) Spec Rep 5: 1–12

Tattersall WM (1926) Crustacea of the order Euphausiacea and Mysidacea from the Western Atlantic. Proc US Natl Mus 69: 1–31

Tattersall WM (1927) Crustaceans of the order Euphausiacea and Mysidacea from the western Atlantic. Proc US Natl Mus 69: 1–28

Tattersall WM (1939) The Euphausiacea and Mysidacea of the John Murray Expedition to the Indian Ocean, 1933-1934. Sci Rep 5: 203–246

Zimmer C (1909) Die Nordischen Schizopoden. Nord Plankton 12: 1–178

Zimmer C (1914) Die Schizopoden der Deutschen Südpolar-Exp, 1901–1903. Dtsch Südpol Exp 15 Zool 7: 377–445

Zimmer C (1915) Schizopoden des Hamburger Naturhistorischen (Zoologischen) Museums. Mitl Naturhist Mus Hamb 32: 159–182

# Appendix

## Stations and Sample data for Euphausiids during R/V *Cariboo* Cruise in the Straits of Magellan

*Euphausia vallentini*

| Station number | Depth (m) | Ind·m$^{-3}$ | Station number | Depth (m) | Ind·m$^{-3}$ | Station number | Depth (m) | Ind·m$^{-3}$ | Station number | Depth (m) | Ind·m$^{-3}$ | Station number | Depth (m) | Ind·m$^{-3}$ | Station number | Depth (m) | Ind·m$^{-3}$ |
|---|---|---|---|---|---|---|---|---|---|---|---|---|---|---|---|---|---|
| 5 | 200–160 | 1.37 | 10 | 600–500 | 0.00 | 14 | 500–400 | 0.02 | 18 | 180–160 | 14.16 | 22 | 30–20 | 0.00 | | | |
|  | 160–140 | 0.37 |  | 500–400 | 0.00 |  | 400–300 | 0.02 |  | 160–120 | 4.45 |  | 20–10 | 0.00 | | | |
|  | 140–115 | 0.02 |  | 400–300 | 0.00 |  | 300–250 | 0.12 |  | 120–100 | 0.35 |  | 10–0 | 0.00 | | | |
|  | 115–100 | 0.21 |  | 300–250 | 0.03 |  | 250–200 | 3.10 |  | 100–80 | 0.19 | | | | | | |
|  | 100–80 | 0.00 |  | 250–200 | 0.00 |  | 200–150 | 21.13 |  | 80–60 | 0.00 | | | | | | |
|  | 80–60 | 0.00 |  | 200–150 | 0.00 |  | 150–100 | 9.73 |  | 60–40 | 0.09 | | | | | | |
|  | 60–40 | 0.00 |  | 150–100 | 0.00 |  | 100–90 | 0.87 |  | 40–20 | 0.06 | | | | | | |
|  | 40–20 | 0.00 |  | 100–50 | 0.00 |  | 90–50 | 0.16 |  | 20–10 | 0.47 | | | | | | |
|  | 20–0 | 0.11 |  | 50–0 | 0.03 |  | 50–0 | 21.01 |  | 10–0 | 0.91 | | | | | | |
| 6 | 500–455 | 0.00 | 11 | 500–400 | 0.01 | 15 | 434–300 | 0.00 | 19 | 106–75 | 0.01 | 23 | 30–20 | 0.00 | | | |
|  | 455–400 | 0.00 |  | 400–300 | 0.00 |  | 300–200 | 0.01 |  | 75–50 | 0.45 |  | 20–10 | 0.00 | | | |
|  | 400–300 | 0.00 |  | 300–250 | 0.01 |  | 200–150 | 0.00 |  | 50–25 | 0.86 |  | 10–0 | 0.00 | | | |
|  | 300–250 | 0.00 |  | 250–200 | 0.00 |  | 150–100 | 0.03 |  | 25–0 | 2.38 | | | | | | |
|  | 250–200 | 0.00 |  | 200–150 | 0.00 |  | 100–80 | 0.08 | | | | | | | | | |
|  | 200–150 | 0.90 |  | 150–100 | 0.00 |  | 80–60 | 0.04 | | | | | | | | | |
|  | 150–100 | 0.19 |  | 100–50 | 0.04 |  | 60–40 | 0.17 | | | | | | | | | |
|  | 100–50 | 0.81 |  | 50–25 | 2.79 |  | 40–20 | 3.85 | | | | | | | | | |
|  | 50–0 | 0.00 |  | 25–0 | 3.19 |  | 20–0 | 29.49 | | | | | | | | | |
| 7 | 600–500 | 0.01 | 12 | 130–100 | 2.47 | 16 | 80–60 | 0.73 | 20 | 80–75 | 0.11 | 26 | 30–20 | 0.00 | | | |
|  | 500–400 | 0.00 |  | 100–50 | 0.01 |  | 60–40 | 0.02 |  | 75–60 | 0.17 |  | 20–10 | 0.00 | | | |
|  | 400–300 | 0.00 |  | 50–25 | 0.03 |  | 40–20 | 0.29 |  | 60–40 | 0.19 |  | 10–0 | 0.00 | | | |
|  | 300–250 | 0.02 |  | 25–0 | 0.03 |  | 20–0 | 4.75 |  | 40–20 | 0.05 | | | | | | |
|  | 250–200 | 0.00 | | | | | | |  | 20–0 | 0.03 | | | | | | |
|  | 200–150 | 3.84 | | | | | | | | | | | | | | | |
|  | 150–100 | 0.89 | | | | | | | | | | | | | | | |
|  | 100–50 | 0.14 | | | | | | | | | | | | | | | |
|  | 50–0 | 0.05 | | | | | | | | | | | | | | | |
| 9 | 900–700 | 0.02 | 13 | 450–400 | 0.00 | 17 | 140–120 | 2.12 | 21 | 40–20 | 0.07 | | | | | | |
|  | 700–600 | 0.00 |  | 400–300 | 0.01 |  | 120–100 | 2.21 |  | 20–0 | 0.02 | | | | | | |
|  | 600–400 | 0.02 |  | 300–250 | 0.00 |  | 100–80 | 0.16 | | | | | | | | | |
|  | 400–300 | 0.00 |  | 250–200 | 0.00 |  | 80–60 | 0.08 | | | | | | | | | |
|  | 300–200 | 0.00 |  | 200–150 | 0.11 |  | 60–40 | 0.00 | | | | | | | | | |
|  | 200–140 | 9.41 |  | 150–100 | 0.75 |  | 40–30 | 0.06 | | | | | | | | | |
|  | 140–100 | 0.91 |  | 100–50 | 0.78 |  | 30–20 | 0.72 | | | | | | | | | |
|  | 100–50 | 0.62 |  | 50–25 | 0.18 |  | 20–10 | 0.00 | | | | | | | | | |
|  | 50–0 | 0.06 |  | 25–0 | 0.00 |  | 10–0 | 0.49 | | | | | | | | | |

*Euphausia lucens*

| Station number | Depth (m) | Ind·m$^{-3}$ | Station number | Depth (m) | Ind·m$^{-3}$ | Station number | Depth (m) | Ind·m$^{-3}$ | Station number | Depth (m) | Ind·m$^{-3}$ | Station number | Depth (m) | Ind·m$^{-3}$ |
|---|---|---|---|---|---|---|---|---|---|---|---|---|---|---|
| 5 | 200–160 | 0.00 | 10 | 600–500 | 0.00 | 14 | 500–400 | 0.00 | 18 | 180–160 | 0.00 | 22 | 30–20 | 0.00 |
|  | 160–140 | 0.02 |  | 500–400 | 0.00 |  | 400–300 | 0.00 |  | 160–120 | 0.00 |  | 20–10 | 0.00 |
|  | 140–115 | 0.00 |  | 400–300 | 0.00 |  | 300–250 | 0.00 |  | 120–100 | 0.00 |  | 10–0 | 0.00 |
|  | 115–100 | 0.00 |  | 300–250 | 0.00 |  | 250–200 | 0.00 |  | 100–80 | 0.00 |  |  |  |
|  | 100–80 | 0.00 |  | 250–200 | 0.00 |  | 200–150 | 0.00 |  | 80–60 | 0.00 |  |  |  |
|  | 80–60 | 0.00 |  | 200–150 | 0.00 |  | 150–100 | 0.00 |  | 60–40 | 0.00 |  |  |  |
|  | 60–40 | 0.00 |  | 150–100 | 0.00 |  | 100–90 | 0.00 |  | 40–20 | 0.00 |  |  |  |
|  | 40–20 | 0.00 |  | 100–50 | 0.00 |  | 90–50 | 0.00 |  | 20–10 | 0.00 |  |  |  |
|  | 20–0 | 0.00 |  | 50–0 | 0.00 |  | 50–0 | 0.00 |  | 10–0 | 0.00 |  |  |  |
| 6 | 500–455 | 0.00 | 11 | 500–400 | 0.00 | 15 | 434–300 | 0.00 | 19 | 106–75 | 0.00 | 23 | 30–20 | 0.00 |
|  | 455–400 | 0.00 |  | 400–300 | 0.00 |  | 300–200 | 0.00 |  | 75–50 | 0.00 |  | 20–10 | 0.00 |
|  | 400–300 | 0.00 |  | 300–250 | 0.00 |  | 200–150 | 0.00 |  | 50–25 | 0.00 |  | 10–0 | 0.00 |
|  | 300–250 | 0.00 |  | 250–200 | 0.00 |  | 150–100 | 0.00 |  | 25–0 | 0.00 |  |  |  |
|  | 250–200 | 0.00 |  | 200–150 | 0.00 |  | 100–80 | 0.00 |  |  |  |  |  |  |
|  | 200–150 | 0.15 |  | 150–100 | 0.00 |  | 80–60 | 0.00 |  |  |  |  |  |  |
|  | 150–100 | 0.19 |  | 100–50 | 0.00 |  | 60–40 | 0.00 |  |  |  |  |  |  |
|  | 100–50 | 0.00 |  | 50–25 | 0.00 |  | 40–20 | 0.00 |  |  |  |  |  |  |
|  | 50–0 | 0.00 |  | 25–0 | 0.00 |  | 20–0 | 0.00 |  |  |  |  |  |  |
| 7 | 600–500 | 0.00 | 12 | 130–100 | 0.00 | 16 | 80–60 | 0.00 | 20 | 80–75 | 0.00 | 26 | 30–20 | 0.07 |
|  | 500–400 | 0.00 |  | 100–50 | 0.00 |  | 60–40 | 0.00 |  | 75–60 | 0.00 |  | 20–10 | 0.00 |
|  | 400–300 | 0.00 |  | 50–25 | 0.00 |  | 40–20 | 0.00 |  | 60–40 | 0.00 |  | 10–0 | 0.00 |
|  | 300–250 | 0.00 |  | 25–0 | 0.00 |  | 20–0 | 0.00 |  | 40–20 | 0.00 |  |  |  |
|  | 250–200 | 0.00 |  |  |  |  |  |  |  | 20–0 | 0.00 |  |  |  |
|  | 200–150 | 0.43 |  |  |  |  |  |  |  |  |  |  |  |  |
|  | 150–100 | 0.16 |  |  |  |  |  |  |  |  |  |  |  |  |
|  | 100–50 | 0.00 |  |  |  |  |  |  |  |  |  |  |  |  |
|  | 50–0 | 0.00 |  |  |  |  |  |  |  |  |  |  |  |  |
| 9 | 900–700 | 0.00 | 13 | 450–400 | 0.00 | 17 | 140–120 | 0.00 | 21 | 40–20 | 0.00 |  |  |  |
|  | 700–600 | 0.00 |  | 400–300 | 0.00 |  | 120–100 | 0.00 |  | 20–0 | 0.00 |  |  |  |
|  | 600–400 | 0.00 |  | 300–250 | 0.00 |  | 100–80 | 0.00 |  |  |  |  |  |  |
|  | 400–300 | 0.00 |  | 250–200 | 0.00 |  | 80–60 | 0.00 |  |  |  |  |  |  |
|  | 300–200 | 0.00 |  | 200–150 | 0.00 |  | 60–40 | 0.00 |  |  |  |  |  |  |
|  | 200–140 | 0.10 |  | 150–100 | 0.00 |  | 40–30 | 0.00 |  |  |  |  |  |  |
|  | 140–100 | 0.68 |  | 100–50 | 0.00 |  | 30–20 | 0.00 |  |  |  |  |  |  |
|  | 100–50 | 0.00 |  | 50–25 | 0.00 |  | 20–10 | 0.00 |  |  |  |  |  |  |
|  | 50–0 | 0.00 |  | 25–0 | 0.00 |  | 10–0 | 0.00 |  |  |  |  |  |  |

*Thysanoessa gregaria*

| Station number | Depth (m) | Ind·m$^{-3}$ | Station number | Depth (m) | Ind·m$^{-3}$ | Station number | Depth (m) | Ind·m$^{-3}$ | Station number | Depth (m) | Ind·m$^{-3}$ | Station number | Depth (m) | Ind·m$^{-3}$ | Station number | Depth (m) | Ind·m$^{-3}$ |
|---|---|---|---|---|---|---|---|---|---|---|---|---|---|---|---|---|---|
| 5 | 200–160 | 0.00 | 10 | 600–500 | 0.00 | 14 | 500–400 | 0.00 | 18 | 180–160 | 0.00 | 22 | 30–20 | 0.00 |
| | 160–140 | 0.00 | | 500–400 | 0.00 | | 400–300 | 0.00 | | 160–120 | 0.00 | | 20–10 | 0.00 |
| | 140–115 | 0.00 | | 400–300 | 0.00 | | 300–250 | 0.00 | | 120–100 | 0.00 | | 10–0 | 0.00 |
| | 115–100 | 0.00 | | 300–250 | 0.00 | | 250–200 | 0.00 | | 100–80 | 0.00 | | | |
| | 100–80 | 0.00 | | 250–200 | 0.00 | | 200–150 | 0.00 | | 80–60 | 0.00 | | | |
| | 80–60 | 0.00 | | 200–150 | 0.00 | | 150–100 | 0.00 | | 60–40 | 0.00 | | | |
| | 60–40 | 0.00 | | 150–100 | 0.00 | | 100–90 | 0.00 | | 40–20 | 0.00 | | | |
| | 40–20 | 0.00 | | 100–50 | 0.00 | | 90–50 | 0.00 | | 20–10 | 0.00 | | | |
| | 20–0 | 0.00 | | 50–0 | 0.00 | | 50–0 | 0.00 | | 10–0 | 0.00 | | | |
| 6 | 500–455 | 0.00 | 11 | 500–400 | 0.00 | 15 | 434–300 | 0.00 | 19 | 106–75 | 0.00 | 23 | 30–20 | 0.00 |
| | 455–400 | 0.00 | | 400–300 | 0.00 | | 300–200 | 0.00 | | 75–50 | 0.00 | | 20–10 | 0.00 |
| | 400–300 | 0.00 | | 300–250 | 0.00 | | 200–150 | 0.00 | | 50–25 | 0.00 | | 10–0 | 0.00 |
| | 300–250 | 0.00 | | 250–200 | 0.00 | | 150–100 | 0.00 | | 25–0 | 0.00 | | | |
| | 250–200 | 0.00 | | 200–150 | 0.00 | | 100–80 | 0.00 | | | | | | |
| | 200–150 | 0.00 | | 150–100 | 0.00 | | 80–60 | 0.00 | | | | | | |
| | 150–100 | 0.00 | | 100–50 | 0.00 | | 60–40 | 0.00 | | | | | | |
| | 100–50 | 0.00 | | 50–25 | 0.00 | | 40–20 | 0.00 | | | | | | |
| | 50–0 | 0.00 | | 25–0 | 0.00 | | 20–0 | 0.00 | | | | | | |
| 7 | 600–500 | 0.00 | 12 | 130–100 | 0.00 | 16 | 80–60 | 0.00 | 20 | 80–75 | 0.00 | 26 | 30–20 | 0.10 |
| | 500–400 | 0.00 | | 100–50 | 0.00 | | 60–40 | 0.00 | | 75–60 | 0.00 | | 20–10 | 0.00 |
| | 400–300 | 0.00 | | 50–25 | 0.00 | | 40–20 | 0.00 | | 60–40 | 0.00 | | 10–0 | 0.00 |
| | 300–250 | 0.00 | | 25–0 | 0.00 | | 20–0 | 0.00 | | 40–20 | 0.00 | | | |
| | 250–200 | 0.00 | | | | | | | | 20–0 | 0.00 | | | |
| | 200–150 | 0.00 | | | | | | | | | | | | |
| | 150–100 | 0.00 | | | | | | | | | | | | |
| | 100–50 | 0.00 | | | | | | | | | | | | |
| | 50–0 | 0.00 | | | | | | | | | | | | |
| 9 | 900–700 | 0.00 | 13 | 450–400 | 0.00 | 17 | 140–120 | 0.00 | 21 | 40–20 | 0.00 | | | |
| | 700–600 | 0.00 | | 400–300 | 0.00 | | 120–100 | 0.00 | | 20–0 | 0.00 | | | |
| | 600–400 | 0.00 | | 300–250 | 0.00 | | 100–80 | 0.00 | | | | | | |
| | 400–300 | 0.00 | | 250–200 | 0.00 | | 80–60 | 0.00 | | | | | | |
| | 300–200 | 0.00 | | 200–150 | 0.00 | | 60–40 | 0.00 | | | | | | |
| | 200–140 | 0.00 | | 150–100 | 0.00 | | 40–30 | 0.00 | | | | | | |
| | 140–100 | 0.00 | | 100–50 | 0.00 | | 30–20 | 0.00 | | | | | | |
| | 100–50 | 0.00 | | 50–25 | 0.00 | | 20–10 | 0.00 | | | | | | |
| | 50–0 | 0.00 | | 25–0 | 0.00 | | 10–0 | 0.00 | | | | | | |

*Nematoscelis megalops*

| Station number | Depth (m) | Ind·m⁻³ | Station number | Depth (m) | Ind·m⁻³ | Station number | Depth (m) | Ind·m⁻³ | Station number | Depth (m) | Ind·m⁻³ | Station number | Depth (m) | Ind·m⁻³ |
|---|---|---|---|---|---|---|---|---|---|---|---|---|---|---|
| 5 | 200–160 | 0.33 | 10 | 600–500 | 0.00 | 14 | 500–400 | 0.00 | 18 | 180–160 | 0.22 | 22 | 30–20 | 0.03 |
|  | 160–140 | 0.17 |  | 500–400 | 0.00 |  | 400–300 | 0.00 |  | 160–120 | 0.61 |  | 20–10 | 0.00 |
|  | 140–115 | 0.19 |  | 400–300 | 0.00 |  | 300–250 | 0.00 |  | 120–100 | 0.13 |  | 10–0 | 0.00 |
|  | 115–100 | 0.18 |  | 300–250 | 0.01 |  | 250–200 | 0.00 |  | 100–80 | 0.05 |  |  |  |
|  | 100–80 | 0.10 |  | 250–200 | 0.00 |  | 200–150 | 0.13 |  | 80–60 | 0.06 |  |  |  |
|  | 80–60 | 0.00 |  | 200–150 | 0.03 |  | 150–100 | 0.32 |  | 60–40 | 0.00 |  |  |  |
|  | 60–40 | 0.70 |  | 150–100 | 0.09 |  | 100–90 | 0.00 |  | 40–20 | 0.00 |  |  |  |
|  | 40–20 | 1.35 |  | 100–50 | 0.00 |  | 90–50 | 0.00 |  | 20–10 | 0.08 |  |  |  |
|  | 20–0 | 0.30 |  | 50–0 | 0.00 |  | 50–0 | 0.57 |  | 10–0 | 0.00 |  |  |  |
| 6 | 500–455 | 0.00 | 11 | 500–400 | 0.00 | 15 | 434–300 | 0.00 | 19 | 106–75 | 0.05 | 23 | 30–20 | 0.00 |
|  | 455–400 | 0.00 |  | 400–300 | 0.01 |  | 300–200 | 0.00 |  | 75–50 | 0.01 |  | 20–10 | 0.00 |
|  | 400–300 | 0.00 |  | 300–250 | 0.00 |  | 200–150 | 0.04 |  | 50–25 | 0.03 |  | 10–0 | 0.00 |
|  | 300–250 | 0.00 |  | 250–200 | 0.04 |  | 150–100 | 0.00 |  | 25–0 | 0.00 |  |  |  |
|  | 250–200 | 0.04 |  | 200–150 | 0.09 |  | 100–80 | 0.00 |  |  |  |  |  |  |
|  | 200–150 | 0.42 |  | 150–100 | 0.12 |  | 80–60 | 0.00 |  |  |  |  |  |  |
|  | 150–100 | 0.07 |  | 100–50 | 0.56 |  | 60–40 | 0.00 |  |  |  |  |  |  |
|  | 100–50 | 0.00 |  | 50–25 | 0.52 |  | 40–20 | 0.00 |  |  |  |  |  |  |
|  | 50–0 | 0.00 |  | 25–0 | 0.13 |  | 20–0 | 0.00 |  |  |  |  |  |  |
| 7 | 600–500 | 0.00 | 12 | 130–100 | 0.06 | 16 | 80–60 | 0.03 | 20 | 80–75 | 0.00 | 26 | 30–20 | 0.03 |
|  | 500–400 | 0.00 |  | 100–50 | 0.00 |  | 60–40 | 0.00 |  | 75–60 | 0.00 |  | 20–10 | 0.00 |
|  | 400–300 | 0.00 |  | 50–25 | 0.00 |  | 40–20 | 0.00 |  | 60–40 | 0.04 |  | 10–0 | 0.00 |
|  | 300–250 | 0.00 |  | 25–0 | 0.00 |  | 20–0 | 0.00 |  | 40–20 | 0.00 |  |  |  |
|  | 250–200 | 0.00 |  |  |  |  |  |  |  | 20–0 | 0.00 |  |  |  |
|  | 200–150 | 0.52 |  |  |  |  |  |  |  |  |  |  |  |  |
|  | 150–100 | 0.08 |  |  |  |  |  |  |  |  |  |  |  |  |
|  | 100–50 | 0.18 |  |  |  |  |  |  |  |  |  |  |  |  |
|  | 50–0 | 0.36 |  |  |  |  |  |  |  |  |  |  |  |  |
| 9 | 900–700 | 0.00 | 13 | 450–400 | 0.00 | 17 | 140–120 | 0.29 | 21 | 40–20 | 0.01 |  |  |  |
|  | 700–600 | 0.00 |  | 400–300 | 0.00 |  | 120–100 | 0.50 |  | 20–0 | 0.00 |  |  |  |
|  | 600–400 | 0.00 |  | 300–250 | 0.00 |  | 100–80 | 0.03 |  |  |  |  |  |  |
|  | 400–300 | 0.00 |  | 250–200 | 0.00 |  | 80–60 | 0.00 |  |  |  |  |  |  |
|  | 300–200 | 0.00 |  | 200–150 | 0.02 |  | 60–40 | 0.00 |  |  |  |  |  |  |
|  | 200–140 | 0.39 |  | 150–100 | 0.10 |  | 40–30 | 0.00 |  |  |  |  |  |  |
|  | 140–100 | 0.14 |  | 100–50 | 0.00 |  | 30–20 | 0.00 |  |  |  |  |  |  |
|  | 100–50 | 1.59 |  | 50–25 | 0.00 |  | 20–10 | 0.00 |  |  |  |  |  |  |
|  | 50–0 | 0.27 |  | 25–0 | 0.00 |  | 10–0 | 0.00 |  |  |  |  |  |  |

*Stylocheiron longicorne*

| Station number | Depth (m) | Ind·m⁻³ | Station number | Depth (m) | Ind·m⁻³ | Station number | Depth (m) | Ind·m⁻³ | Station number | Depth (m) | Ind·m⁻³ | Station number | Depth (m) | Ind·m⁻³ |
|---|---|---|---|---|---|---|---|---|---|---|---|---|---|---|
| 5 | 200–160 | 0.00 | 10 | 600–500 | 0.00 | 14 | 500–400 | 0.00 | 18 | 180–160 | 0.00 | 22 | 30–20 | 0.00 |
|  | 160–140 | 0.00 |  | 500–400 | 0.00 |  | 400–300 | 0.00 |  | 160–120 | 0.00 |  | 20–10 | 0.00 |
|  | 140–115 | 0.00 |  | 400–300 | 0.00 |  | 300–250 | 0.00 |  | 120–100 | 0.00 |  | 10–0 | 0.00 |
|  | 115–100 | 0.00 |  | 300–250 | 0.00 |  | 250–200 | 0.00 |  | 100–80 | 0.00 |  |  |  |
|  | 100–80 | 0.00 |  | 250–200 | 0.00 |  | 200–150 | 0.00 |  | 80–60 | 0.00 |  |  |  |
|  | 80–60 | 0.00 |  | 200–150 | 0.00 |  | 150–100 | 0.03 |  | 60–40 | 0.00 |  |  |  |
|  | 60–40 | 0.00 |  | 150–100 | 0.00 |  | 100–90 | 0.00 |  | 40–20 | 0.00 |  |  |  |
|  | 40–20 | 0.00 |  | 100–50 | 0.00 |  | 90–50 | 0.00 |  | 20–10 | 0.00 |  |  |  |
|  | 20–0 | 0.00 |  | 50–0 | 0.00 |  | 50–0 | 0.00 |  | 10–0 | 0.00 |  |  |  |
| 6 | 500–455 | 0.00 | 11 | 500–400 | 0.00 | 15 | 434–300 | 0.00 | 19 | 106–75 | 0.00 | 23 | 30–20 | 0.00 |
|  | 455–400 | 0.00 |  | 400–300 | 0.00 |  | 300–200 | 0.00 |  | 75–50 | 0.00 |  | 20–10 | 0.00 |
|  | 400–300 | 0.00 |  | 300–250 | 0.00 |  | 200–150 | 0.00 |  | 50–25 | 0.00 |  | 10–0 | 0.00 |
|  | 300–250 | 0.00 |  | 250–200 | 0.00 |  | 150–100 | 0.00 |  | 25–0 | 0.00 |  |  |  |
|  | 250–200 | 0.00 |  | 200–150 | 0.00 |  | 100–80 | 0.00 |  |  |  |  |  |  |
|  | 200–150 | 0.00 |  | 150–100 | 0.00 |  | 80–60 | 0.00 |  |  |  |  |  |  |
|  | 150–100 | 0.00 |  | 100–50 | 0.00 |  | 60–40 | 0.00 |  |  |  |  |  |  |
|  | 100–50 | 0.00 |  | 50–25 | 0.00 |  | 40–20 | 0.00 |  |  |  |  |  |  |
|  | 50–0 | 0.00 |  | 25–0 | 0.00 |  | 20–0 | 0.00 |  |  |  |  |  |  |
| 7 | 600–500 | 0.00 | 12 | 130–100 | 0.00 | 16 | 80–60 | 0.00 | 20 | 80–75 | 0.00 | 26 | 30–20 | 0.00 |
|  | 500–400 | 0.00 |  | 100–50 | 0.00 |  | 60–40 | 0.00 |  | 75–60 | 0.00 |  | 20–10 | 0.00 |
|  | 400–300 | 0.00 |  | 50–25 | 0.00 |  | 40–20 | 0.00 |  | 60–40 | 0.00 |  | 10–0 | 0.00 |
|  | 300–250 | 0.00 |  | 25–0 | 0.00 |  | 20–0 | 0.00 |  | 40–20 | 0.00 |  |  |  |
|  | 250–200 | 0.00 |  |  |  |  |  |  |  | 20–0 | 0.00 |  |  |  |
|  | 200–150 | 0.00 |  |  |  |  |  |  |  |  |  |  |  |  |
|  | 150–100 | 0.00 |  |  |  |  |  |  |  |  |  |  |  |  |
|  | 100–50 | 0.00 |  |  |  |  |  |  |  |  |  |  |  |  |
|  | 50–0 | 0.00 |  |  |  |  |  |  |  |  |  |  |  |  |
| 9 | 900–700 | 0.00 | 13 | 450–400 | 0.00 | 17 | 140–120 | 0.00 | 21 | 40–20 | 0.00 |  |  |  |
|  | 700–600 | 0.00 |  | 400–300 | 0.00 |  | 120–100 | 0.00 |  | 20–0 | 0.00 |  |  |  |
|  | 600–400 | 0.00 |  | 300–250 | 0.00 |  | 100–80 | 0.00 |  |  |  |  |  |  |
|  | 400–300 | 0.00 |  | 250–200 | 0.00 |  | 80–60 | 0.00 |  |  |  |  |  |  |
|  | 300–200 | 0.00 |  | 200–150 | 0.00 |  | 60–40 | 0.00 |  |  |  |  |  |  |
|  | 200–140 | 0.00 |  | 150–100 | 0.00 |  | 40–30 | 0.00 |  |  |  |  |  |  |
|  | 140–100 | 0.00 |  | 100–50 | 0.00 |  | 30–20 | 0.00 |  |  |  |  |  |  |
|  | 100–50 | 0.03 |  | 50–25 | 0.00 |  | 20–10 | 0.00 |  |  |  |  |  |  |
|  | 50–0 | 0.00 |  | 25–0 | 0.00 |  | 10–0 | 0.00 |  |  |  |  |  |  |

*Stylocheiron maximum*

| Station number | Depth (m) | Ind·m⁻³ | Station number | Depth (m) | Ind·m⁻³ | Station number | Depth (m) | Ind·m⁻³ | Station number | Depth (m) | Ind·m⁻³ | Station number | Depth (m) | Ind·m⁻³ |
|---|---|---|---|---|---|---|---|---|---|---|---|---|---|---|
| 5 | 200–160 | 0.00 | 10 | 600–500 | 0.00 | 14 | 500–400 | 0.00 | 18 | 180–160 | 0.00 | 22 | 30–20 | 0.00 |
|   | 160–140 | 0.00 |    | 500–400 | 0.00 |    | 400–300 | 0.00 |    | 160–120 | 0.00 |    | 20–10 | 0.00 |
|   | 140–115 | 0.00 |    | 400–300 | 0.00 |    | 300–250 | 0.00 |    | 120–100 | 0.00 |    | 10–0 | 0.00 |
|   | 115–100 | 0.00 |    | 300–250 | 0.00 |    | 250–200 | 0.00 |    | 100–80 | 0.00 |    |   |   |
|   | 100–80 | 0.00 |    | 250–200 | 0.00 |    | 200–150 | 0.00 |    | 80–60 | 0.00 |    |   |   |
|   | 80–60 | 0.00 |    | 200–150 | 0.00 |    | 150–100 | 0.00 |    | 60–40 | 0.00 |    |   |   |
|   | 60–40 | 0.00 |    | 150–100 | 0.00 |    | 100–90 | 0.00 |    | 40–20 | 0.00 |    |   |   |
|   | 40–20 | 0.00 |    | 100–50 | 0.00 |    | 90–50 | 0.00 |    | 20–10 | 0.00 |    |   |   |
|   | 20–0 | 0.00 |    | 50–0 | 0.00 |    | 50–0 | 0.00 |    | 10–0 | 0.00 |    |   |   |
| 6 | 500–455 | 0.00 | 11 | 500–400 | 0.00 | 15 | 434–300 | 0.00 | 19 | 106–75 | 0.00 | 23 | 30–20 | 0.00 |
|   | 455–400 | 0.00 |    | 400–300 | 0.00 |    | 300–200 | 0.01 |    | 75–50 | 0.00 |    | 20–10 | 0.00 |
|   | 400–300 | 0.00 |    | 300–250 | 0.00 |    | 200–150 | 0.00 |    | 50–25 | 0.00 |    | 10–0 | 0.00 |
|   | 300–250 | 0.00 |    | 250–200 | 0.00 |    | 150–100 | 0.00 |    | 25–0 | 0.00 |    |   |   |
|   | 250–200 | 0.00 |    | 200–150 | 0.00 |    | 100–80 | 0.00 |    |   |   |    |   |   |
|   | 200–150 | 0.00 |    | 150–100 | 0.00 |    | 80–60 | 0.00 |    |   |   |    |   |   |
|   | 150–100 | 0.00 |    | 100–50 | 0.00 |    | 60–40 | 0.00 |    |   |   |    |   |   |
|   | 100–50 | 0.00 |    | 50–25 | 0.00 |    | 40–20 | 0.00 |    |   |   |    |   |   |
|   | 50–0 | 0.00 |    | 25–0 | 0.00 |    | 20–0 | 0.00 |    |   |   |    |   |   |
| 7 | 600–500 | 0.00 | 12 | 130–100 | 0.00 | 16 | 80–60 | 0.00 | 20 | 80–75 | 0.00 | 26 | 30–20 | 0.00 |
|   | 500–400 | 0.00 |    | 100–50 | 0.00 |    | 60–40 | 0.00 |    | 75–60 | 0.00 |    | 20–10 | 0.00 |
|   | 400–300 | 0.00 |    | 50–25 | 0.00 |    | 40–20 | 0.00 |    | 60–40 | 0.00 |    | 10–0 | 0.00 |
|   | 300–250 | 0.00 |    | 25–0 | 0.00 |    | 20–0 | 0.00 |    | 40–20 | 0.00 |    |   |   |
|   | 250–200 | 0.00 |    |   |   |    |   |   |    | 20–0 | 0.00 |    |   |   |
|   | 200–150 | 0.00 |    |   |   |    |   |   |    |   |   |    |   |   |
|   | 150–100 | 0.00 |    |   |   |    |   |   |    |   |   |    |   |   |
|   | 100–50 | 0.00 |    |   |   |    |   |   |    |   |   |    |   |   |
|   | 50–0 | 0.00 |    |   |   |    |   |   |    |   |   |    |   |   |
| 9 | 900–700 | 0.00 | 13 | 450–400 | 0.00 | 17 | 140–120 | 0.00 | 21 | 40–20 | 0.00 |   |   |   |
|   | 700–600 | 0.00 |    | 400–300 | 0.00 |    | 120–100 | 0.00 |    | 20–0 | 0.00 |    |   |   |
|   | 600–400 | 0.00 |    | 300–250 | 0.00 |    | 100–80 | 0.00 |    |   |   |    |   |   |
|   | 400–300 | 0.00 |    | 250–200 | 0.00 |    | 80–60 | 0.00 |    |   |   |    |   |   |
|   | 300–200 | 0.00 |    | 200–150 | 0.00 |    | 60–40 | 0.00 |    |   |   |    |   |   |
|   | 200–140 | 0.00 |    | 150–100 | 0.00 |    | 40–30 | 0.00 |    |   |   |    |   |   |
|   | 140–100 | 0.00 |    | 100–50 | 0.00 |    | 30–20 | 0.00 |    |   |   |    |   |   |
|   | 100–50 | 0.00 |    | 50–25 | 0.00 |    | 20–10 | 0.00 |    |   |   |    |   |   |
|   | 50–0 | 0.00 |    | 25–0 | 0.00 |    | 10–0 | 0.00 |    |   |   |    |   |   |

# Chapter 5 Mysids

N. Crescenti

## Systematic Account

Subphylum CRUSTACEA
Class MALACOSTRACA

Order MYSIDACEA
Suborder MYSIDA

**Family Mysidae**
Subfamily Boreomysinae

## 5.1 *Boreomysis rostrata* Illig, 1906

*Boreomysis rostrata* Illig, 1906: pp. 196, 197, fig.
2a–d – Illig, 1930: pp. 414–419, figs. 23–27 – Tattersall, 1951: pp. 56–60, figs. 11-13 – Tattersall, 1955: pp. 68–71 – Holmquist, 1957: pp. 5–10, fig. 1a–t – Ii, 1964: pp. 20-23 – Müller, 1993: p. 36 (world catalogue).
*Boreomysis inermis* Hansen, 1910: pp. 26, 27, pl. 2, fig 4a–c.
*Boreomysis rostrata japonica* Ii, 1964: pp. 23–28, figs. 1, 2.
*Boreomysis rostrata orientalis* Ii, 1964: pp. 28–33, figs. 3, 4.

Body length –
females: 18.5–20.8 mm (30 specimens)
males: 18.2–20.3 mm (30 specimens)

**Female.** Rostrum of carapace pointed and elongated, reaching distal margin of eyes. Lateral margins of frontal plate convex and covering basal parts of ocular peduncles. Eyes bearing conspicuous ocular papilla. Antennules with small 2nd article of peduncle, 1/2 as long as wide. Distal margin of 3rd segment with small tubercle bearing several fine setules. Antennal scale about 4 times longer than wide, surpassing distal margin of antennular peduncle by ca. 1/3. Outer margin unarmed and terminating distally with tooth; apex of scale slightly surpassing tooth. Uropods long and slender; exopod longer than endopod by 1/4; both are longer than telson. Exopods unarmed for short region in proximal area of outer margins. This region terminates with 2 small spines of unequal length. Endopod bearing long, slender and recurved spine on dorsal margin below statocysts. Telson elongated and cleft; proximally wide and distally

narrow. Proximal 1/4 of lateral margins unarmed whereas remaining 3/4 bearing series of long, regularly distributed spines, with smaller spines in between. Cleft narrow and deep, with characteristic dilation at base. Margins of cleft armed with dense series of spinules, closely aposed to one another.

**Male.** Similar to female except for the general differences in the pleopods and a tuft of long setules inserted on the antennule, at the base of the 2 flagella.

**Remarks**

I agree with the description given by Tattersall (1955) and Holmquist (1956, 1957). There are no striking differences between females and males except for the pleopods, which are more developed in males and bear a thick tuft of setules at the base of the flagella of the antennular peduncle. The length of the rostrum is more or less the same in both sexes; the specimens examined in the present study had a somewhat longer rostrum, reaching the tip of the eyes in both females and males.

**Distribution**

According to Holmquist (1957), the species is confined to the Southern Oceans, south of the Tropic of Capricorn. It seems to be a deep-dwelling species, and was sampled from 245–470 m during the Lund University Chile Expedition and as deep as 2000 m during the *Valdivia* Expedition. In the present study, it was the only species of mysid frequently sampled in the Straits, especially in the Pacific sector, with very few specimens collected in the Central area (Sts. 16 and 17).

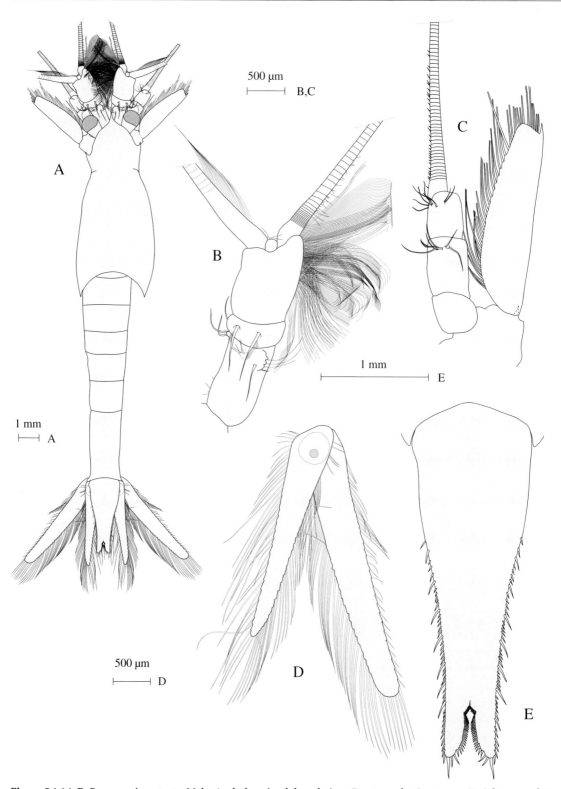

**Figure 5.1.1A-E.** *Boreomysis rostrata*. Male: **A** whole animal dorsal view; **B** antennule; **C** antenna; **D** right uropod; **E** telson

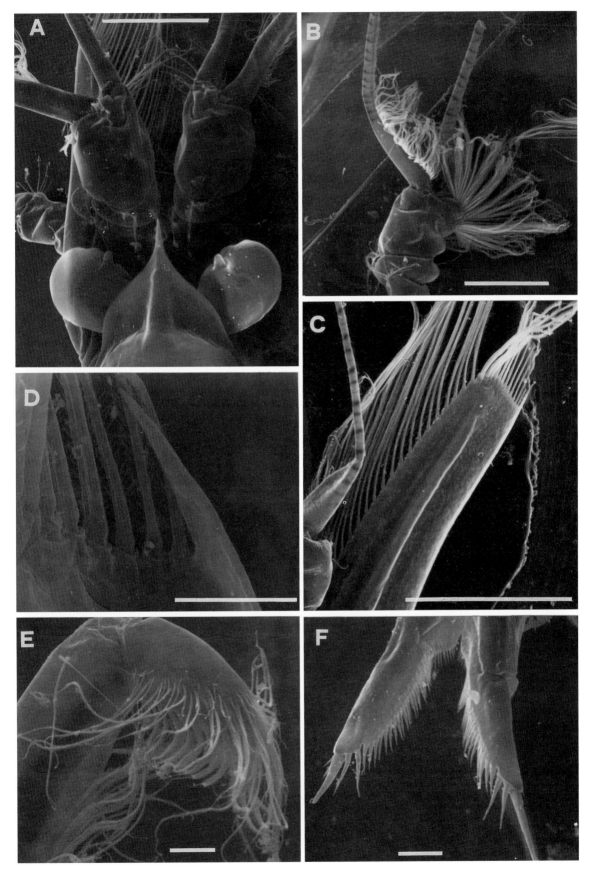

**Fig. 5.1.2A-F.** *Boreomysis rostrata*. Male: **A** anterior end of the carapace; **B** antennular peduncle; **C** antennal scale; **D** detail of tooth of antennal scale; **E** distal end of endopod of 2nd thoracic appendage; **F** telson cleft. *Bars* **A,B,C** 1 mm; **D,E,F** 100 μm

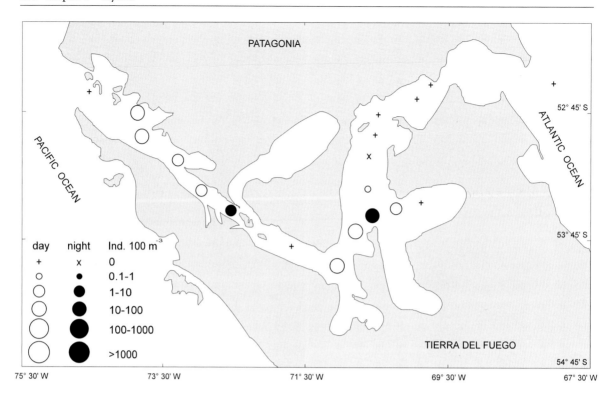

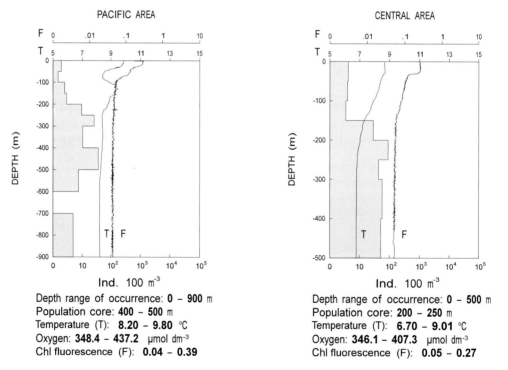

Depth range of occurrence: **0 – 900** m
Population core: **400 – 500** m
Temperature (T): **8.20 – 9.80** °C
Oxygen: **348.4 – 437.2** μmol dm-3
Chl fluorescence (F): **0.04 – 0.39**

Depth range of occurrence: **0 – 500** m
Population core: **200 – 250** m
Temperature (T): **6.70 – 9.01** °C
Oxygen: **346.1 – 407.3** μmol dm-3
Chl fluorescence (F): **0.05 – 0.27**

**Fig. 5.1.3.** *Boreomysis rostrata.* Distribution in the Straits of Magellan in February-March 1991

## 5.2 *Pseudomma magellanensis* Tattersall, 1955

*Pseudomma magellanensis* Tattersall, 1955: pp. 100, 101 – Mauchline and Murano, 1977: p. 74 (world catalogue) – Müller, 1993: p. 137 (world catalogue)

Body length –
   female: 8.8 mm (1 specimen)
   male: 8.4 mm (1 specimen)

**Female.** Carapace with short anterior margin; slightly convex and uncovering the high plate. 3rd segment of antennule longer than wide and equal in length to ca. sum of the first 2 segments. Dorsal surface at base of 2 flagellae bearing rectangular tubercle with robust spine on anterior outer margin. Inner margin bearing 2 very small spines and 2 slender setules. Antennal scale small, 3 times longer than wide, surpassing distal margin of antennular peduncle by 1/3. Unarmed outer margin terminating with wide tooth. Apex of scale slightly surpassing tooth margin. Wide ocular plate, typical of the genus, projecting forward to reach proximal limit of 2nd segment of antennular peduncle; not entirely convex but bearing marked bulges along median line of 2 halves. Rudimentary eyes appear contiguous. Lateral margins of plate armed with 9 irregular teeth extending along lateral margins by 3/4, beginning from antero-lateral corners. Uropod with small pointed endopod extending beyond telson by 1/2. Exopod wider and much longer, reaching beyond distal margin of endopod by about 1/4. Telson short and with concave lateral margins, armed with 8 spines at distal 2/3 of margin; these progressively increase in size distally. Apex wide and convex, armed with 3 long slender spines that become progressively longer towards median line. Here, a pair of plumose setules are inserted on a small papilla.

**Male.** Ventral side of distal margin of antennular peduncle bearing typical lobe with thick tuft of long setules. 4th pleopod with 8-segmented exopod, longer than endopod. Endopod 7-segmented. All pleopods bearing long and slender simpodites and thread-like rami.

### Remarks

*P. magellanensis* is similar to the congener *P. sarsi* but can be distinguished by its smaller size and sunken ocular plate with 2 well-defined dorsal crests separated by a shallow depression (Tattersall 1955). Also, the species has smaller amorphous teeth along the lateral margins of the ocular plate, the 4th pleopod of the males differs, and the telson has a characteristic apex with concave lateral margins. Differences have been observed in the number of setules on the rectangular tubercle at the base of the 2 antennular flagella. Males have only 2 setules, whereas females possess 4. This is probably due to the great individual variability observed in species of the genus *Pseudomma*, as already noted by O.S. Tattersall (1955) while studying the material from the *Discovery* collection.

### Distribution

To date, this species had been reported only by Tattersall (1955), who recorded a single female and male specimen in night samples collected from 0–300 m in the Straits of Magellan. Tattersall assigned the specimens to a new species. In the present study, the only 2 specimens were sampled from 0–100 m at St. 19 in the Central area of the Straits.

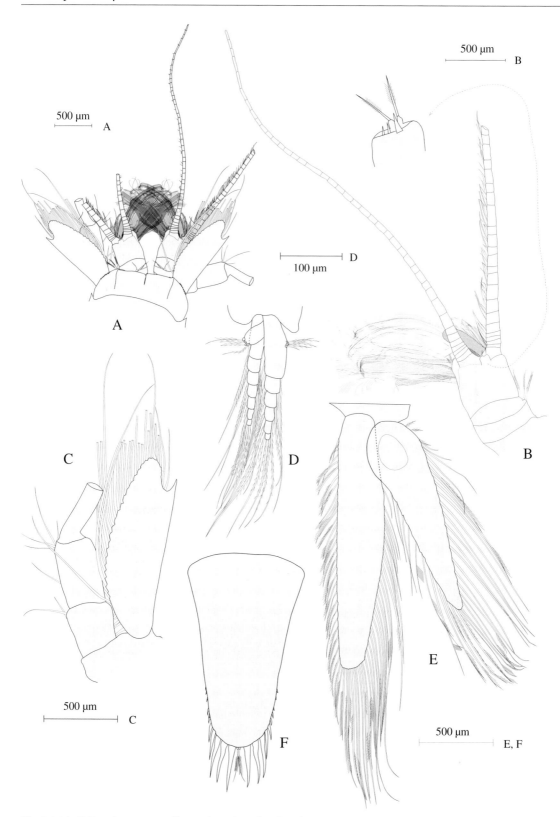

**Fig. 5.2.1A–F.** *Pseudomma magellanensis.* Male: **A** head with antennae 1 and 2; **B** antennule; **C** antenna; **D** 4th pleopod; **E** left uropod; **F** telson

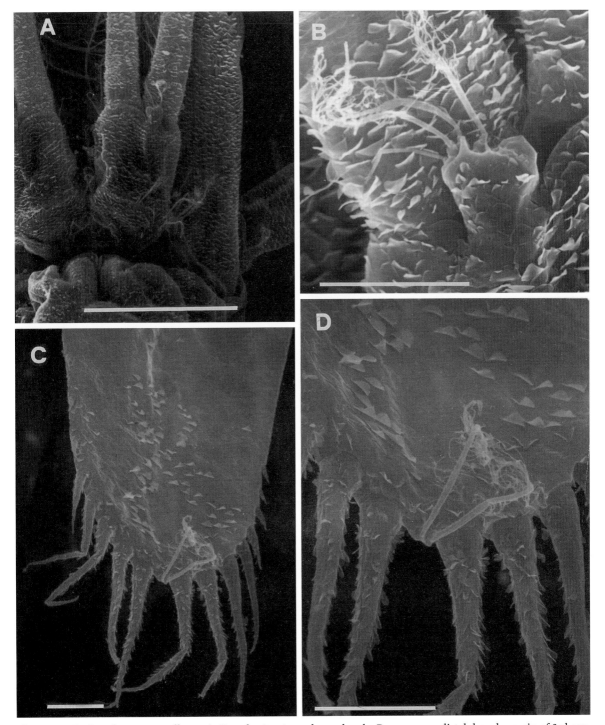

**Fig. 5.2.2A–D.** *Pseudomma magellanensis.* Female: **A** antennular peduncle; **B** process on distal dorsal margin of 3rd segment of antennular peduncle; **C** end of telson; **D** detail of pair of plumose setae on apex of telson. *Bars* **A** 500 μm; **B,C,E** 100 μm

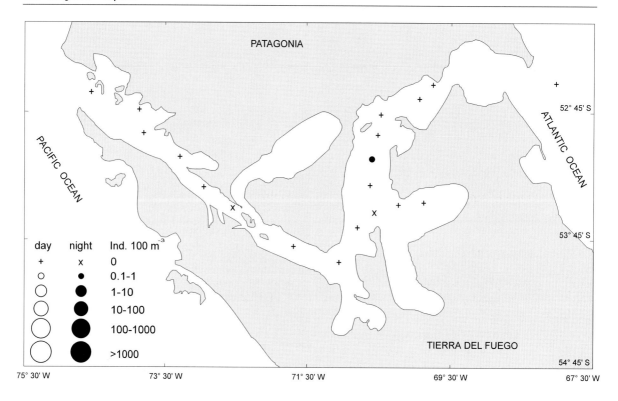

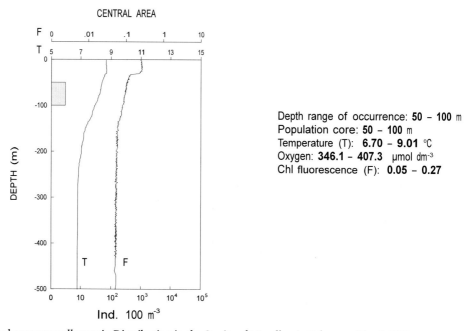

Depth range of occurrence: **50 – 100** m
Population core: **50 – 100** m
Temperature (T): **6.70 – 9.01** °C
Oxygen: **346.1 – 407.3** μmol dm⁻³
Chl fluorescence (F): **0.05 – 0.27**

**Fig. 5.2.3.** *Pseudomma magellanensis.* Distribution in the Straits of Magellan in February-March 1991

## 5.3 *Arthromysis magellanica* Cunningham, 1871

*Macromysis magellanica* Cunningham, 1871: p. 497.
*Antarctomysis sp.* Coutière, 1906: pp. 1–10, 2 pls.
*Arthromysis chierchiae* Colosi, 1924: pp. 3–5, figs. 1–3.
*Arthromysis magellanica* Tattersall, 1955: pp. 179–181, figs. 45, 46 – Mauchline and Murano, 1977: p. 48 (world catalogue) – Müller, 1993: p. 212 (world catalogue).

Body length –
immature females: 12.0–13.5 mm
(8 specimens)

**Immature female.** Posterior margin of carapace short, uncovering the 8th, 7th and part of the 6th thoracic somites. Anterior portion with rounded, semicircular frontal plate. Antero-lateral corners of carapace spine-like. Eyes with elongated, slender and cylindrical peduncle, with hemispherical cornea. Eyes longer than distal extremity of 2nd segment of antennular peduncle. Antennules bearing elongated peduncle. 3rd segment almost as long as 2nd, with expanded distal extremity. A small nodule is evident at base of flagellae, bearing 4 slender setules and spiniform process. Antenna with long and slender scale, twice the length of antennular peduncle; antenna bearing setae on both sides and with distal suture. Pleon with 6th segment twice as long as 5th. Uropod with long and slender exopod, twice as long as telson. Endopod shorter and about 2/3 as long as exopod; densely covered with setae and bearing series of 13 spines on inner side, in submarginal position, below statocysts. Telson 3 times longer than wide. Margins

of telson bearing series of 24–25 regularly distributed spines. Terminal spines closely apposed to shorter apical spines, giving rise to spinulation on telson cleft. There are about 45 of these short spines on each side. 2 long and plumose setules initiate from inner side of cleft and extend beyond distal end by twice their length.

### Remarks

This species, referred to as *Arthromysis chierchiae* by Colosi (1924), who erected the genus, posed several systematic problems until Tattersall (1955) examined a sufficient number of specimens to attribute them to *A. magellanica*. The species is characterized by two long and plumose setules arising from the base of telson cleft. Even though the species can be easily confused with *Tenagomysis tenuipes*, it can be distinguished by its elongated, slender eyes and semicircular form of the anterior margin of the carapace (Tattersall 1955).

### Distribution

This species was first reported by Cunningham (1871) at the eastern end of the Straits of Magellan. It was later also reported in the Straits by Zimmer (1915). Colosi (1924) found a single female specimen at Cape Virgin (Straits of Magellan), whereas Tattersall (1955) recorded numerous adult specimens in the area comprised between the Falkland Islands and Tierra del Fuego, at a depth of 21–23 m. Tattersall (op. cit.) also sampled it in the Straits at a depth of 0–40 m. In the present study, *A. magellanica* was sampled only from 10–30 m at St. 26, in Atlantic waters beyond the Straits.

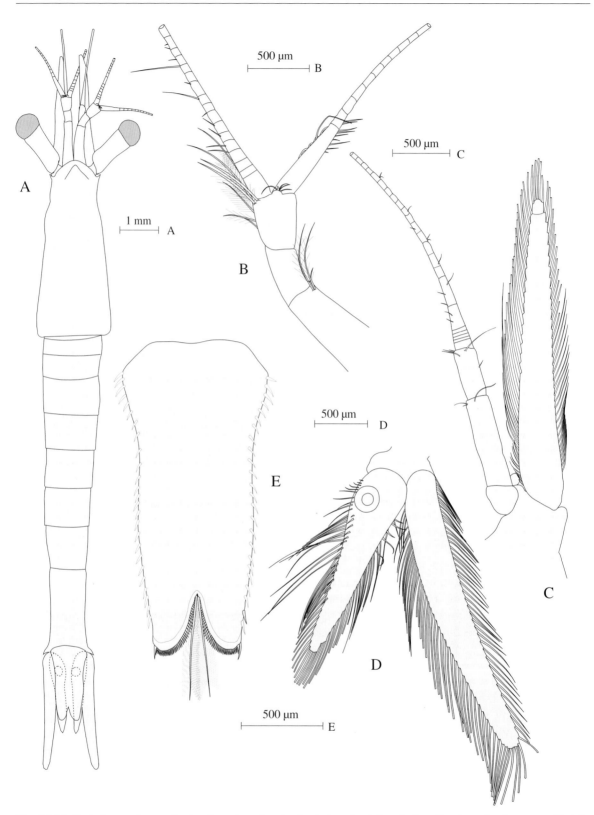

**Fig. 5.3.1A–E.** *Arthromysis magellanica*. Immature female: **A** whole animal dorsal view; **B** antennule; **C** antenna; **D** right uropod; **E** telson

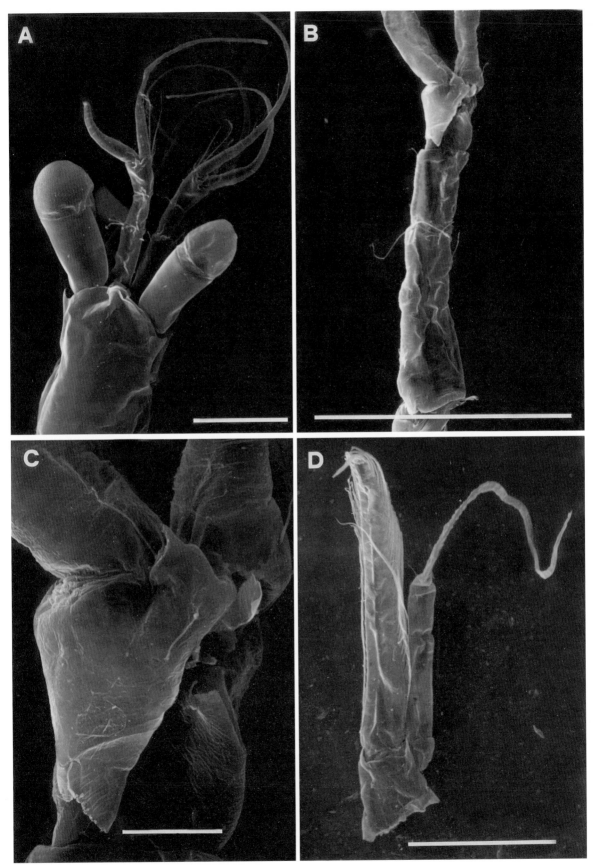

**Fig. 5.3.2A–D.** *Arthromysis magellanica.* Female: **A** anterior end of carapace; **B** antennular peduncle; **C** detail of small nodule at base of antennular flagellae; **D** antenna. *Bars* **A,B,D** 1 mm; **C** 100 μm

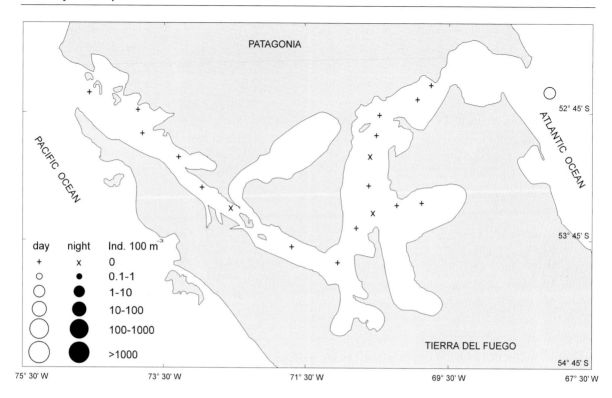

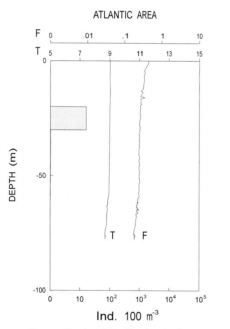

Depth range of occurrence: **0 – 30** m
Population core: **20 – 30** m
Temperature (T): **8.68 – 9.02** °C
Oxygen: **390.4 – 410.9** μmol dm[-3]
Chl fluorescence (F): **0.16 – 0.44**

**Fig. 5.3.3.** *Arthromysis magellanica.* Distribution in the Straits of Magellan in February-March 1991

## 5.4 *Neomysis monticellii* Colosi, 1924

*Neomysis monticellii* Colosi, 1924: pp. 6, 7, figs. 7–9 – Tattersall, 1951: p. 248 – Tattersall, 1955: pp. 171, 172, figs. 43a, 44a–c – Müller, 1993: p. 240 (world catalogue).

Body length –
   females: 6.4–10.5 mm (11 specimens)
   males: 6.5 mm (13 specimens)

**Female.** Body small; posterodorsal margin of carapace reaching anterior edge of 8th thoracic somite. Anteriorly, dorsal plate bearing very short, triangular frontal plate, not covering eyes. Apex of rostrum slightly rounded and with acute extremity, bending downwards at the base of ocular peduncles. Eyes large and darkly pigmented. Antennules with typical 3-segmented peduncle; 2nd segment 1/3 length of last. Antennae with relatively large peduncle, surpassing extremity of antennular peduncles. Scale lanceolated, both margins of which covered with setules; length of scale extending beyond antennular peduncles by 1/5. Pleon with 6th segment 1/3 longer than 5th. Uropodal endopodite somewhat longer than telson; exopod longer than telson by 1/3. Ventral side of endopodite bearing 2 spines, in correspondence to distal margin of statocysts. Of the 2, shorter spine on external margin. Telson about 1.5 times longer than last somite of pleon, elongated, entire, and with pointed extremity. Along margins of either side are 30-32 equal or subequal spines; extremity bears 2 pairs of spines. 1st pair of more outer spines are longer than spines on margins; 2nd pair of inner spines shorter than 1st pair and those on margins.

**Male.** Antennules bearing characteristic protuberance on ventral side at base of 2 flagellae. Protuberance well developed and as long as last segment of antennular peduncle; lobate and bearing thick tuft of long setules. Antennal scale reaching apex of lobe. 4th pleopod long; exopod 4 times longer than endopod, extending somewhat beyond apex of telson. Exopod consisting in two segments; 1st segment almost 4 times longer than 2nd. Same pleopod bearing 2 large spinulated setules of unequal length at distal extremity. Endopod with numerous setules on inner margin; outer margin bearing small tubercle with 4 setae, bending over exopod.

### Remarks

The protuberance on the distal ventral side of the antennular peduncle of the male is very stout, lobate and bristly. Colosi (1924) reported that a distinct distal articulation was visible on the scale of the antenna. However, in accordance with Tattersall (1955), this was not observed in the present study. My observations also agree with Tattersall and not with Colosi with regards to the 2 spines on the ventral side of the uropodal endopodite. The latter author reported the presence of a single spine in this position. The spines observed on the outer margins of the telson ranged from 30–32 as compared to 23–28 reported by Colosi. Tattersall does not make any mention of the number of these spines.

### Distribution.

*N. monticellii* has formerly been reported only by Colosi (1924) and Tattersall (1955). The former author found the species at Cape Virgin (Straits of Magellan) but did not report the depth of sampling. Tattersall described 3 specimens, all males, from the *Discovery* collection. These had been sampled in coastal waters at the NE extremity of the Straits of Magellan at a depth of 49–66 m. In the present study, all of the specimens were collected in the vicinity of the Straits, but not within the Straits proper, in Atlantic waters at a depth of 10–30 m.

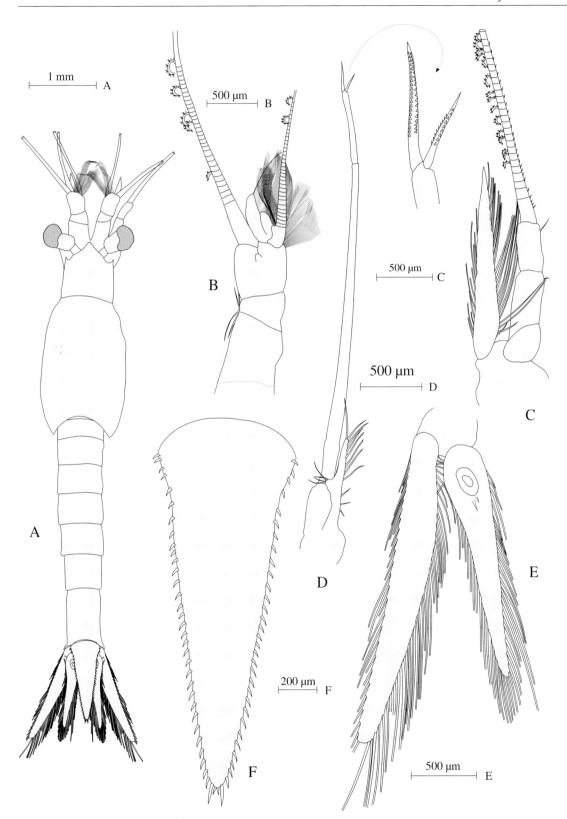

**Fig. 5.4.1A–F.** *Neomysis monticellii.* Male: **A** whole animal dorsal view; **B** antennule; **C** antenna; **D** 4th pleopod; **E** left uropod; **F** telson

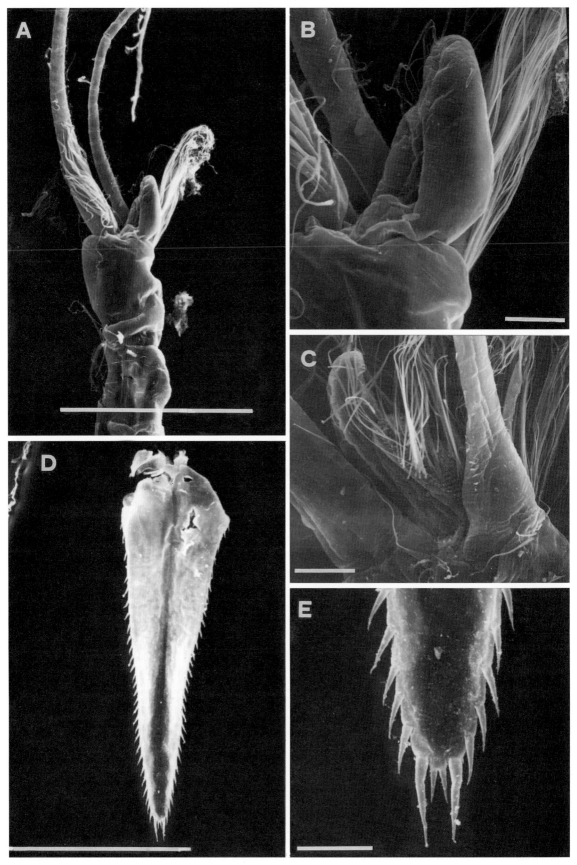

**Fig. 5.4.2A–E.** *Neomysis monticellii*. Male: **A** antennular peduncle; **B** detail of antennular protuberance in laterodorsal view; **C** detail of thick tuft of long setules on antennular protuberance in ventral view; **D** telson in dorsal view; **E** detail of terminal end of telson. *Bars* **A,D** 1 mm; **B,C,E** 100 µm

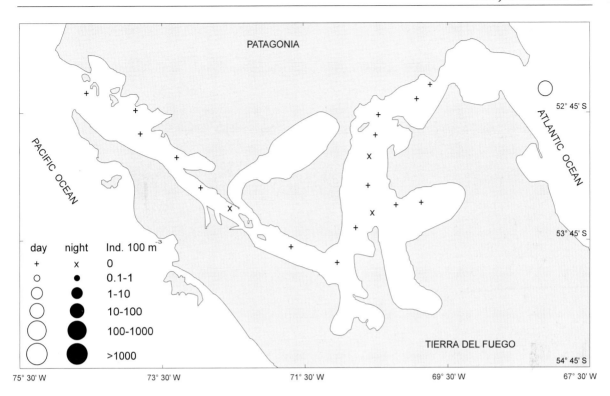

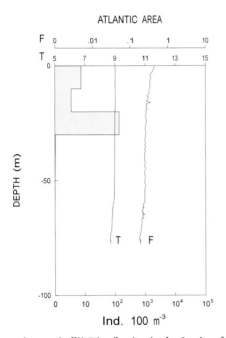

ATLANTIC AREA

Depth range of occurrence: **0 – 30** m
Population core: **20 – 30** m
Temperature (T): **8.68 – 9.02** °C
Oxygen: **390.4 – 410.9** µmol dm⁻³
Chl fluorescence (F): **0.16 – 0.44**

**Fig. 5.4.3.** *Neomysis monticellii*. Distribution in the Straits of Magellan in February-March 1991

# References

Colosi G (1924) Euphausiacea e Mysidacea raccolti dalla R. Nave *Vettor Pisani* nel 1882–1885. Ann Mus Zool Univ Napoli New Ser 5: 1–7

Coutière MH (1906) Crustacés schizopodes et décapodes. J Charcot Exped Antarct Fr (1903–1905): 1–10

Cunningham RO (1871) Notes on the reptiles, Amphibia, Fishes, Mollusca, and Crustacea obtained during the voyage of HMS *Nassau* in the years 1866–1869. Trans Linn Soc Lond 1, 27: 491–502, 2 pls

Hansen HJ (1910) The Schizopoda of the Siboga Expedition. Siboga Exped 37: 1–123, 16 pls

Holmquist C (1956) Betrachtungen über *Boreomysis rostrata* Illig und weitere *Boreomysis* Arten. Ark Zool 2, 10: 427–447

Holmquist C (1957) Mysidacea of Chile. Reports of the Lund University of Chile Expedition 1948–1949 (Report 28). Lunds Univ Arsskr NF Adv 2, 53: 1–52

Ii N (1964) Fauna Japonica, Mysidae. Biogeogr Soc Jpn Tokyo, pp. 610

Illig G (1906) Bericht über die neuen Schizopodengattungen und -arten der deutschen Tiefsee Expedition 1898–1899. Zool Anz 30: 194–211

Illig G (1930) Die Schizopoden der deutschen Tiefsee Expedition. Rep Valdivia Exped 22: 397–625

Mauchline J, Murano M (1977) World list of the Mysidacea, Crustacea. J Tokyo Univ Fish 64: 39–88

Müller HG (1993) World catalogue and bibliography of the recent Mysidacea. Verlag HG Müller, Laboratory for Tropical Ecosystems Research & Information Service, Wetzlar 491 pp

Tattersall OS (1955) Mysidacea. Discovery Rep 28: 1–190

Tattersall WM (1951) A review of the Mysidacea of the United States National Museum. Smithson Inst US Natl Mus Bull 201: 1–292

Zimmer C (1915) Schizopoden des Hamburger Naturhistorischen (Zoologischen) Museums. Mitt Naturhist Mus Hamb 32: 159–182

## Stations and Sample Data for Mysids During R/V *Cariboo* Cruise in the Straits of Magellan

*Boreomysis rostrata*

| Station number | Depth (m) | Ind·m$^{-3}$ | Station number | Depth (m) | Ind·m$^{-3}$ | Station number | Depth (m) | Ind·m$^{-3}$ | Station number | Depth (m) | Ind·m$^{-3}$ | Station number | Depth (m) | Ind·m$^{-3}$ |
|---|---|---|---|---|---|---|---|---|---|---|---|---|---|---|
| 5 | 200–160 | 0.00 | 10 | 600–500 | 0.00 | 14 | 500–400 | 0.35 | 18 | 180–160 | 0.02 | 22 | 30–20 | 0.00 |
|  | 160–140 | 0.00 |  | 500–400 | 0.01 |  | 400–300 | 0.32 |  | 160–120 | 0.00 |  | 20–10 | 0.00 |
|  | 140–115 | 0.00 |  | 400–300 | 0.00 |  | 300–250 | 0.54 |  | 120–100 | 0.00 |  | 10–0 | 0.00 |
|  | 115–100 | 0.00 |  | 300–250 | 0.01 |  | 250–200 | 0.76 |  | 100–80 | 0.00 |  |  |  |
|  | 100–80 | 0.00 |  | 250–200 | 0.01 |  | 200–150 | 0.06 |  | 80–60 | 0.00 |  |  |  |
|  | 80–60 | 0.00 |  | 200–150 | 0.08 |  | 150–100 | 0.00 |  | 60–40 | 0.00 |  |  |  |
|  | 60–40 | 0.00 |  | 150–100 | 0.05 |  | 100–90 | 0.00 |  | 40–20 | 0.00 |  |  |  |
|  | 40–20 | 0.00 |  | 100–50 | 0.00 |  | 90–50 | 0.00 |  | 20–10 | 0.00 |  |  |  |
|  | 20–0 | 0.00 |  | 50–0 | 0.00 |  | 50–0 | 0.00 |  | 10–0 | 0.00 |  |  |  |
| 6 | 500–455 | 0.75 | 11 | 500–400 | 0.11 | 15 | 434–300 | 0.69 | 19 | 106–75 | 0.00 | 23 | 30–20 | 0.00 |
|  | 455–400 | 0.39 |  | 400–300 | 0.04 |  | 300–200 | 0.34 |  | 75–50 | 0.00 |  | 20–10 | 0.00 |
|  | 400–300 | 0.13 |  | 300–250 | 0.00 |  | 200–150 | 0.44 |  | 50–25 | 0.00 |  | 10–0 | 0.00 |
|  | 300–250 | 0.30 |  | 250–200 | 0.01 |  | 150–100 | 0.13 |  | 25–0 | 0.00 |  |  |  |
|  | 250–200 | 0.04 |  | 200–150 | 0.00 |  | 100–80 | 0.22 |  |  |  |  |  |  |
|  | 200–150 | 0.00 |  | 150–100 | 0.01 |  | 80–60 | 0.04 |  |  |  |  |  |  |
|  | 150–100 | 0.00 |  | 100–50 | 0.03 |  | 60–40 | 0.09 |  |  |  |  |  |  |
|  | 100–50 | 0.00 |  | 50–25 | 0.04 |  | 40–20 | 0.04 |  |  |  |  |  |  |
|  | 50–0 | 0.00 |  | 25–0 | 0.00 |  | 20–0 | 0.00 |  |  |  |  |  |  |
| 7 | 600–500 | 0.13 | 12 | 130–100 | 0.00 | 16 | 80–60 | 0.00 | 20 | 80–75 | 0.00 | 26 | 30–20 | 0.00 |
|  | 500–400 | 0.49 |  | 100–50 | 0.00 |  | 60–40 | 0.00 |  | 75–60 | 0.00 |  | 20–10 | 0.00 |
|  | 400–300 | 0.24 |  | 50–25 | 0.00 |  | 40–20 | 0.12 |  | 60–40 | 0.00 |  | 10–0 | 0.00 |
|  | 300–250 | 0.66 |  | 25–0 | 0.00 |  | 20–0 | 0.00 |  | 40–20 | 0.00 |  |  |  |
|  | 250–200 | 0.27 |  |  |  |  |  |  |  | 20–0 | 0.00 |  |  |  |
|  | 200–150 | 0.02 |  |  |  |  |  |  |  |  |  |  |  |  |
|  | 150–100 | 0.00 |  |  |  |  |  |  |  |  |  |  |  |  |
|  | 100–50 | 0.00 |  |  |  |  |  |  |  |  |  |  |  |  |
|  | 50–0 | 0.02 |  |  |  |  |  |  |  |  |  |  |  |  |
| 9 | 900–700 | 0.04 | 13 | 450–400 | 0.66 | 17 | 140–120 | 0.00 | 21 | 40–20 | 0.00 |  |  |  |
|  | 700–600 | 0.00 |  | 400–300 | 0.68 |  | 120–100 | 0.00 |  | 20–0 | 0.00 |  |  |  |
|  | 600–400 | 0.07 |  | 300–250 | 0.31 |  | 100–80 | 0.00 |  |  |  |  |  |  |
|  | 400–300 | 0.06 |  | 250–200 | 1.03 |  | 80–60 | 0.00 |  |  |  |  |  |  |
|  | 300–200 | 0.18 |  | 200–150 | 0.59 |  | 60–40 | 0.00 |  |  |  |  |  |  |
|  | 200–140 | 0.00 |  | 150–100 | 0.00 |  | 40–30 | 0.00 |  |  |  |  |  |  |
|  | 140–100 | 0.05 |  | 100–50 | 0.00 |  | 30–20 | 0.00 |  |  |  |  |  |  |
|  | 100–50 | 0.00 |  | 50–25 | 0.09 |  | 20–10 | 0.00 |  |  |  |  |  |  |
|  | 50–0 | 0.00 |  | 25–0 | 0.03 |  | 10–0 | 0.00 |  |  |  |  |  |  |

*Pseudomma magellanensis*

| Station number | Depth (m) | Ind·m$^{-3}$ | Station number | Depth (m) | Ind·m$^{-3}$ | Station number | Depth (m) | Ind·m$^{-3}$ | Station number | Depth (m) | Ind·m$^{-3}$ | Station number | Depth (m) | Ind·m$^{-3}$ |
|---|---|---|---|---|---|---|---|---|---|---|---|---|---|---|
| 5 | 200–160 | 0.00 | 10 | 600–500 | 0.00 | 14 | 500–400 | 0.00 | 18 | 180–160 | 0.00 | 22 | 30–20 | 0.00 |
|   | 160–140 | 0.00 |    | 500–400 | 0.00 |    | 400–300 | 0.00 |    | 160–120 | 0.00 |    | 20–10 | 0.00 |
|   | 140–115 | 0.00 |    | 400–300 | 0.00 |    | 300–250 | 0.00 |    | 120–100 | 0.00 |    | 10–0 | 0.00 |
|   | 115–100 | 0.00 |    | 300–250 | 0.00 |    | 250–200 | 0.00 |    | 100–80 | 0.00 |    |   |   |
|   | 100–80 | 0.00 |    | 250–200 | 0.00 |    | 200–150 | 0.00 |    | 80–60 | 0.00 |    |   |   |
|   | 80–60 | 0.00 |    | 200–150 | 0.00 |    | 150–100 | 0.00 |    | 60–40 | 0.00 |    |   |   |
|   | 60–40 | 0.00 |    | 150–100 | 0.00 |    | 100–90 | 0.00 |    | 40–20 | 0.00 |    |   |   |
|   | 40–20 | 0.00 |    | 100–50 | 0.00 |    | 90–50 | 0.00 |    | 20–10 | 0.00 |    |   |   |
|   | 20–0 | 0.00 |    | 50–0 | 0.00 |    | 50–0 | 0.00 |    | 10–0 | 0.00 |    |   |   |
| 6 | 500–455 | 0.00 | 11 | 500–400 | 0.00 | 15 | 434–300 | 0.00 | 19 | 106–75 | 0.02 | 23 | 30–20 | 0.00 |
|   | 455–400 | 0.00 |    | 400–300 | 0.00 |    | 300–200 | 0.00 |    | 75–50 | 0.00 |    | 20–10 | 0.00 |
|   | 400–300 | 0.00 |    | 300–250 | 0.00 |    | 200–150 | 0.00 |    | 50–25 | 0.00 |    | 10–0 | 0.00 |
|   | 300–250 | 0.00 |    | 250–200 | 0.00 |    | 150–100 | 0.00 |    | 25–0 | 0.00 |    |   |   |
|   | 250–200 | 0.00 |    | 200–150 | 0.00 |    | 100–80 | 0.00 |    |   |   |    |   |   |
|   | 200–150 | 0.00 |    | 150–100 | 0.00 |    | 80–60 | 0.00 |    |   |   |    |   |   |
|   | 150–100 | 0.00 |    | 100–50 | 0.00 |    | 60–40 | 0.00 |    |   |   |    |   |   |
|   | 100–50 | 0.00 |    | 50–25 | 0.00 |    | 40–20 | 0.00 |    |   |   |    |   |   |
|   | 50–0 | 0.00 |    | 25–0 | 0.00 |    | 20–0 | 0.00 |    |   |   |    |   |   |
| 7 | 600–500 | 0.00 | 12 | 130–100 | 0.00 | 16 | 80–60 | 0.00 | 20 | 80–75 | 0.00 | 26 | 30–20 | 0.00 |
|   | 500–400 | 0.00 |    | 100–50 | 0.00 |    | 60–40 | 0.00 |    | 75–60 | 0.00 |    | 20–10 | 0.00 |
|   | 400–300 | 0.00 |    | 50–25 | 0.00 |    | 40–20 | 0.00 |    | 60–40 | 0.00 |    | 10–0 | 0.00 |
|   | 300–250 | 0.00 |    | 25–0 | 0.00 |    | 20–0 | 0.00 |    | 40–20 | 0.00 |    |   |   |
|   | 250–200 | 0.00 |    |   |   |    |   |   |    | 20–0 | 0.00 |    |   |   |
|   | 200–150 | 0.00 |    |   |   |    |   |   |    |   |   |    |   |   |
|   | 150–100 | 0.00 |    |   |   |    |   |   |    |   |   |    |   |   |
|   | 100–50 | 0.00 |    |   |   |    |   |   |    |   |   |    |   |   |
|   | 50–0 | 0.00 |    |   |   |    |   |   |    |   |   |    |   |   |
| 9 | 900–700 | 0.00 | 13 | 450–400 | 0.00 | 17 | 140–120 | 0.00 | 21 | 40–20 | 0.00 |   |   |   |
|   | 700–600 | 0.00 |    | 400–300 | 0.00 |    | 120–100 | 0.00 |    | 20–0 | 0.00 |    |   |   |
|   | 600–400 | 0.00 |    | 300–250 | 0.00 |    | 100–80 | 0.00 |    |   |   |    |   |   |
|   | 400–300 | 0.00 |    | 250–200 | 0.00 |    | 80–60 | 0.00 |    |   |   |    |   |   |
|   | 300–200 | 0.00 |    | 200–150 | 0.00 |    | 60–40 | 0.00 |    |   |   |    |   |   |
|   | 200–140 | 0.00 |    | 150–100 | 0.00 |    | 40–30 | 0.00 |    |   |   |    |   |   |
|   | 140–100 | 0.00 |    | 100–50 | 0.00 |    | 30–20 | 0.00 |    |   |   |    |   |   |
|   | 100–50 | 0.00 |    | 50–25 | 0.00 |    | 20–10 | 0.00 |    |   |   |    |   |   |
|   | 50–0 | 0.00 |    | 25–0 | 0.00 |    | 10–0 | 0.00 |    |   |   |    |   |   |

*Artromysis magellanica*

| Station number | Depth (m) | Ind·m$^{-3}$ | Station number | Depth (m) | Ind·m$^{-3}$ | Station number | Depth (m) | Ind·m$^{-3}$ | Station number | Depth (m) | Ind·m$^{-3}$ | Station number | Depth (m) | Ind·m$^{-3}$ |
|---|---|---|---|---|---|---|---|---|---|---|---|---|---|---|
| 5 | 200–160 | 0.00 | 10 | 600–500 | 0.00 | 14 | 500–400 | 0.00 | 18 | 180–160 | 0.00 | 22 | 30–20 | 0.00 |
|   | 160–140 | 0.00 |    | 500–400 | 0.00 |    | 400–300 | 0.00 |    | 160–120 | 0.00 |    | 20–10 | 0.00 |
|   | 140–115 | 0.00 |    | 400–300 | 0.00 |    | 300–250 | 0.00 |    | 120–100 | 0.00 |    | 10–0 | 0.00 |
|   | 115–100 | 0.00 |    | 300–250 | 0.00 |    | 250–200 | 0.00 |    | 100–80 | 0.00 |    |    |    |
|   | 100–80 | 0.00 |    | 250–200 | 0.00 |    | 200–150 | 0.00 |    | 80–60 | 0.00 |    |    |    |
|   | 80–60 | 0.00 |    | 200–150 | 0.00 |    | 150–100 | 0.00 |    | 60–40 | 0.00 |    |    |    |
|   | 60–40 | 0.00 |    | 150–100 | 0.00 |    | 100–90 | 0.00 |    | 40–20 | 0.00 |    |    |    |
|   | 40–20 | 0.00 |    | 100–50 | 0.00 |    | 90–50 | 0.00 |    | 20–10 | 0.00 |    |    |    |
|   | 20–0 | 0.00 |    | 50–0 | 0.00 |    | 50–0 | 0.00 |    | 10–0 | 0.00 |    |    |    |
| 6 | 500–455 | 0.00 | 11 | 500–400 | 0.00 | 15 | 434–300 | 0.00 | 19 | 106–75 | 0.00 | 23 | 30–20 | 0.00 |
|   | 455–400 | 0.00 |    | 400–300 | 0.00 |    | 300–200 | 0.00 |    | 75–50 | 0.00 |    | 20–10 | 0.00 |
|   | 400–300 | 0.00 |    | 300–250 | 0.00 |    | 200–150 | 0.00 |    | 50–25 | 0.00 |    | 10–0 | 0.00 |
|   | 300–250 | 0.00 |    | 250–200 | 0.00 |    | 150–100 | 0.00 |    | 25–0 | 0.00 |    |    |    |
|   | 250–200 | 0.00 |    | 200–150 | 0.00 |    | 100–80 | 0.00 |    |    |    |    |    |    |
|   | 200–150 | 0.00 |    | 150–100 | 0.00 |    | 80–60 | 0.00 |    |    |    |    |    |    |
|   | 150–100 | 0.00 |    | 100–50 | 0.00 |    | 60–40 | 0.00 |    |    |    |    |    |    |
|   | 100–50 | 0.00 |    | 50–25 | 0.00 |    | 40–20 | 0.00 |    |    |    |    |    |    |
|   | 50–0 | 0.00 |    | 25–0 | 0.00 |    | 20–0 | 0.00 |    |    |    |    |    |    |
| 7 | 600–500 | 0.00 | 12 | 130–100 | 0.00 | 16 | 80–60 | 0.00 | 20 | 80–75 | 0.00 | 26 | 30–20 | 0.15 |
|   | 500–400 | 0.00 |    | 100–50 | 0.00 |    | 60–40 | 0.00 |    | 75–60 | 0.00 |    | 20–10 | 0.00 |
|   | 400–300 | 0.00 |    | 50–25 | 0.00 |    | 40–20 | 0.00 |    | 60–40 | 0.00 |    | 10–0 | 0.00 |
|   | 300–250 | 0.00 |    | 25–0 | 0.00 |    | 20–0 | 0.00 |    | 40–20 | 0.00 |    |    |    |
|   | 250–200 | 0.00 |    |    |    |    |    |    |    | 20–0 | 0.00 |    |    |    |
|   | 200–150 | 0.00 |    |    |    |    |    |    |    |    |    |    |    |    |
|   | 150–100 | 0.00 |    |    |    |    |    |    |    |    |    |    |    |    |
|   | 100–50 | 0.00 |    |    |    |    |    |    |    |    |    |    |    |    |
|   | 50–0 | 0.00 |    |    |    |    |    |    |    |    |    |    |    |    |
| 9 | 900–700 | 0.00 | 13 | 450–400 | 0.00 | 17 | 140–120 | 0.00 | 21 | 40–20 | 0.00 |    |    |    |
|   | 700–600 | 0.00 |    | 400–300 | 0.00 |    | 120–100 | 0.00 |    | 20–0 | 0.00 |    |    |    |
|   | 600–400 | 0.00 |    | 300–250 | 0.00 |    | 100–80 | 0.00 |    |    |    |    |    |    |
|   | 400–300 | 0.00 |    | 250–200 | 0.00 |    | 80–60 | 0.00 |    |    |    |    |    |    |
|   | 300–200 | 0.00 |    | 200–150 | 0.00 |    | 60–40 | 0.00 |    |    |    |    |    |    |
|   | 200–140 | 0.00 |    | 150–100 | 0.00 |    | 40–30 | 0.00 |    |    |    |    |    |    |
|   | 140–100 | 0.00 |    | 100–50 | 0.00 |    | 30–20 | 0.00 |    |    |    |    |    |    |
|   | 100–50 | 0.00 |    | 50–25 | 0.00 |    | 20–10 | 0.00 |    |    |    |    |    |    |
|   | 50–0 | 0.00 |    | 25–0 | 0.00 |    | 10–0 | 0.00 |    |    |    |    |    |    |

*Neomysis monticelli*

| Station number | Depth (m) | Ind·m⁻³ | Station number | Depth (m) | Ind·m⁻³ | Station number | Depth (m) | Ind·m⁻³ | Station number | Depth (m) | Ind·m⁻³ | Station number | Depth (m) | Ind·m⁻³ |
|---|---|---|---|---|---|---|---|---|---|---|---|---|---|---|
| 5 | 200–160 | 0.00 | 10 | 600–500 | 0.00 | 14 | 500–400 | 0.00 | 18 | 180–160 | 0.00 | 22 | 30–20 | 0.00 |
|  | 160–140 | 0.00 |  | 500–400 | 0.00 |  | 400–300 | 0.00 |  | 160–120 | 0.00 |  | 20–10 | 0.00 |
|  | 140–115 | 0.00 |  | 400–300 | 0.00 |  | 300–250 | 0.00 |  | 120–100 | 0.00 |  | 10–0 | 0.00 |
|  | 115–100 | 0.00 |  | 300–250 | 0.00 |  | 250–200 | 0.00 |  | 100–80 | 0.00 |  |  |  |
|  | 100–80 | 0.00 |  | 250–200 | 0.00 |  | 200–150 | 0.00 |  | 80–60 | 0.00 |  |  |  |
|  | 80–60 | 0.00 |  | 200–150 | 0.00 |  | 150–100 | 0.00 |  | 60–40 | 0.00 |  |  |  |
|  | 60–40 | 0.00 |  | 150–100 | 0.00 |  | 100–90 | 0.00 |  | 40–20 | 0.00 |  |  |  |
|  | 40–20 | 0.00 |  | 100–50 | 0.00 |  | 90–50 | 0.00 |  | 20–10 | 0.00 |  |  |  |
|  | 20–0 | 0.00 |  | 50–0 | 0.00 |  | 50–0 | 0.00 |  | 10–0 | 0.00 |  |  |  |
| 6 | 500–455 | 0.00 | 11 | 500–400 | 0.00 | 15 | 434–300 | 0.00 | 19 | 106–75 | 0.00 | 23 | 30–20 | 0.00 |
|  | 455–400 | 0.00 |  | 400–300 | 0.00 |  | 300–200 | 0.00 |  | 75–50 | 0.00 |  | 20–10 | 0.00 |
|  | 400–300 | 0.00 |  | 300–250 | 0.00 |  | 200–150 | 0.00 |  | 50–25 | 0.00 |  | 10–0 | 0.00 |
|  | 300–250 | 0.00 |  | 250–200 | 0.00 |  | 150–100 | 0.00 |  | 25–0 | 0.00 |  |  |  |
|  | 250–200 | 0.00 |  | 200–150 | 0.00 |  | 100–80 | 0.00 |  |  |  |  |  |  |
|  | 200–150 | 0.00 |  | 150–100 | 0.00 |  | 80–60 | 0.00 |  |  |  |  |  |  |
|  | 150–100 | 0.00 |  | 100–50 | 0.00 |  | 60–40 | 0.00 |  |  |  |  |  |  |
|  | 100–50 | 0.00 |  | 50–25 | 0.00 |  | 40–20 | 0.00 |  |  |  |  |  |  |
|  | 50–0 | 0.00 |  | 25–0 | 0.00 |  | 20–0 | 0.00 |  |  |  |  |  |  |
| 7 | 600–500 | 0.00 | 12 | 130–100 | 0.00 | 16 | 80–60 | 0.00 | 20 | 80–75 | 0.00 | 26 | 30–20 | 1.24 |
|  | 500–400 | 0.00 |  | 100–50 | 0.00 |  | 60–40 | 0.00 |  | 75–60 | 0.00 |  | 20–10 | 0.03 |
|  | 400–300 | 0.00 |  | 50–25 | 0.00 |  | 40–20 | 0.00 |  | 60–40 | 0.00 |  | 10–0 | 0.05 |
|  | 300–250 | 0.00 |  | 25–0 | 0.00 |  | 20–0 | 0.00 |  | 40–20 | 0.00 |  |  |  |
|  | 250–200 | 0.00 |  |  |  |  |  |  |  | 20–0 | 0.00 |  |  |  |
|  | 200–150 | 0.00 |  |  |  |  |  |  |  |  |  |  |  |  |
|  | 150–100 | 0.00 |  |  |  |  |  |  |  |  |  |  |  |  |
|  | 100–50 | 0.00 |  |  |  |  |  |  |  |  |  |  |  |  |
|  | 50–0 | 0.00 |  |  |  |  |  |  |  |  |  |  |  |  |
| 9 | 900–700 | 0.00 | 13 | 450–400 | 0.00 | 17 | 140–120 | 0.00 | 21 | 40–20 | 0.00 |  |  |  |
|  | 700–600 | 0.00 |  | 400–300 | 0.00 |  | 120–100 | 0.00 |  | 20–0 | 0.00 |  |  |  |
|  | 600–400 | 0.00 |  | 300–250 | 0.00 |  | 100–80 | 0.00 |  |  |  |  |  |  |
|  | 400–300 | 0.00 |  | 250–200 | 0.00 |  | 80–60 | 0.00 |  |  |  |  |  |  |
|  | 300–200 | 0.00 |  | 200–150 | 0.00 |  | 60–40 | 0.00 |  |  |  |  |  |  |
|  | 200–140 | 0.00 |  | 150–100 | 0.00 |  | 40–30 | 0.00 |  |  |  |  |  |  |
|  | 140–100 | 0.00 |  | 100–50 | 0.00 |  | 30–20 | 0.00 |  |  |  |  |  |  |
|  | 100–50 | 0.00 |  | 50–25 | 0.00 |  | 20–10 | 0.00 |  |  |  |  |  |  |
|  | 50–0 | 0.00 |  | 25–0 | 0.00 |  | 10–0 | 0.00 |  |  |  |  |  |  |

# Chapter 6  Ostracods

K. G. McKenzie, G. Benassi, and I. Ferrari

## Systematic Account

## 6.1 *Mikroconchoecia* cf. *acuticosta* Müller, 1906

*Conchoecia acuticosta* Müller, 1906a: p. 87, pl. 30 figs. 18–21 – Angel, 1981: p. 556, fig. 194/15.
*Microconchoecia acuticosta* Poulsen, 1973: p. 67.
*Conchoecia (Microconchoecia) acuticosta* Deevey, 1978: p. 69, 70, fig. 14 – Deevey, 1982: p. 156, fig. 27.
*Mikroconchoecia* aff. *acuticosta* Martens, 1979: p. 349.

Carapace length –
   females: 0.98–1.26 mm
   males: 1.10–1.30 mm

**Female.** Carapace rectangular, squat and short; the height:length ratio is about 0.65; rostrum slightly curved; incisure broad and moderately deep; dorsal margin relatively straight, apart from weakly developed shoulders; posterior margin convex, posterodorsally each valve ends in a barely discernible blunted tip; ventral margin likewise convex and smoothly curved. Shell sculpture (ornament) distinct, consisting of very thin parallel riblets posteriorly and ventrally, becoming reticulate toward the centre of each valve and anteriorly; the intersection of this meshwork with the anteroventral margin produces a denticulate effect. Both asymmetric glands in their usual places, i.e. postero-dorsal in the left valve and near postero-ventral in the right valve; the right gland is large and prominent (this is a good differentiating character between *acuticosta* and the type species *M. curta*). Cap of the frontal organ short and smooth, united with but somewhat thicker than the stem. All limbs relatively short. A1 indistinctly segmented in front; of the 5 end bristles, the 'e' bristle is about as long as the segments combined, 3 of the sensory 'pipe' bris-

tles are shorter and deeply bifurcate, the 4th is simple but about the same length. Furca with 8 claws separated from the lamella and reducing in size rearwards; there is no unpaired posterior bristle.

**Male.** Carapace similar to that of female. Cap of the frontal organ separated from the stem by a distinct joint, broadly rounded in front on our specimens and smooth. A1 with 5 end bristles; the 'e' bristle ornamented with a single series of 12 spinules; of the 4 sensory 'pipe' bristles, 2 are slender and nearly as long as the chief bristle, a 3rd is shorter and deeply bifurcate, and the last is simple and short. The clasping joints of the A2 are dimorphic; the right one is relatively slender and strongly recurved with a retrousse tip (unlike *M. curta*), the left one is short and almost straight. Penis rather short and stout, with somewhat sinuous margins and a broadly rounded front; a few transverse muscle bands; seminal tube simple.

### Remarks

Only a few specimens were found, in samples from five Magellan Stations (7, 16, 18, 20, 21). We follow Martens (1979) in assigning this species with some hesitation. Our adults are smaller than the size range given by Angel (1981). On the other hand, Martens (1979), Angel (1981) and Deevey (1983) all recorded *M. curta* (Lubbock 1860) – our first determination for this form – as only occurring north of 47°S; while Martens (1979) identified *M. acuticosta* as the more austral representative of this rather similar pair of species. Martens (1979) also noted that the carapace sculpture is very variable, even in specimens from the same haul, and sometimes is almost invisible.

## Distribution

Angel (1981) regards this as a mesopelagic? and transoceanic species which occurs most frequently in high latitudes. In the Atlantic, its previously known range was from 60°N to 45°S (Angel 1981). Poulsen (1973) notes that it was taken in the southern part of the Indian and west Pacific Oceans at 20–100 m depth. Deevey (1978, fig. 14) records it from the south Pacific between about 90–160°W and from 33–54°S. Deevey (1982) adds numerous Southern and Indian Ocean records between about 115–165°E and from 35–50°S; and gives its Indian Ocean latitudinal range as from 5°N to 50°S, noting that the species seems to be restricted to the upper few hundred metres. On the other hand, Hartmann (1985) gives the depth range as 0–2274 m. Assuming that his deep water records are not contaminants, it may be that the species moves with Antarctic Bottom Water into more northerly (lower) latitudes, where it is taken at depth.

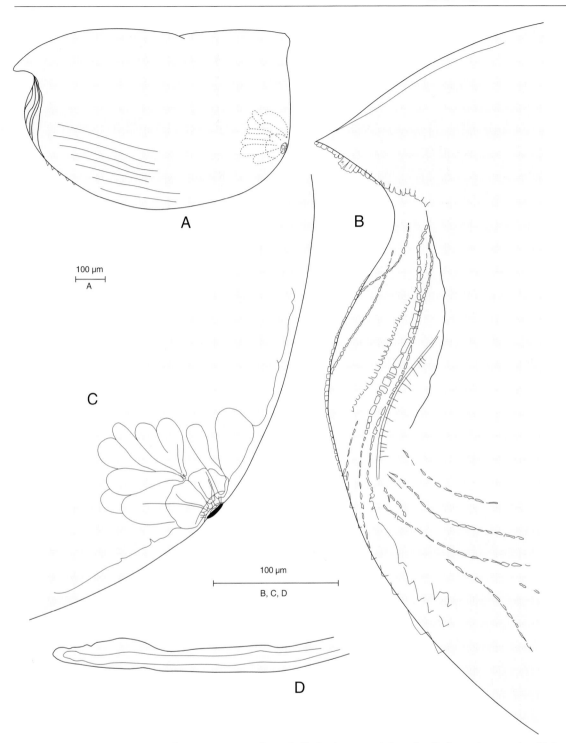

100 µm
A

100 µm
B, C, D

**Fig. 6.1.1A–D.** *Mikroconchoecia* cf. *acuticosta*. Female: **A** shell of ovigerous adult; **B** detail of anterior margin of left valve; **C** detail of posterior gland of right valve; **D** frontal organ

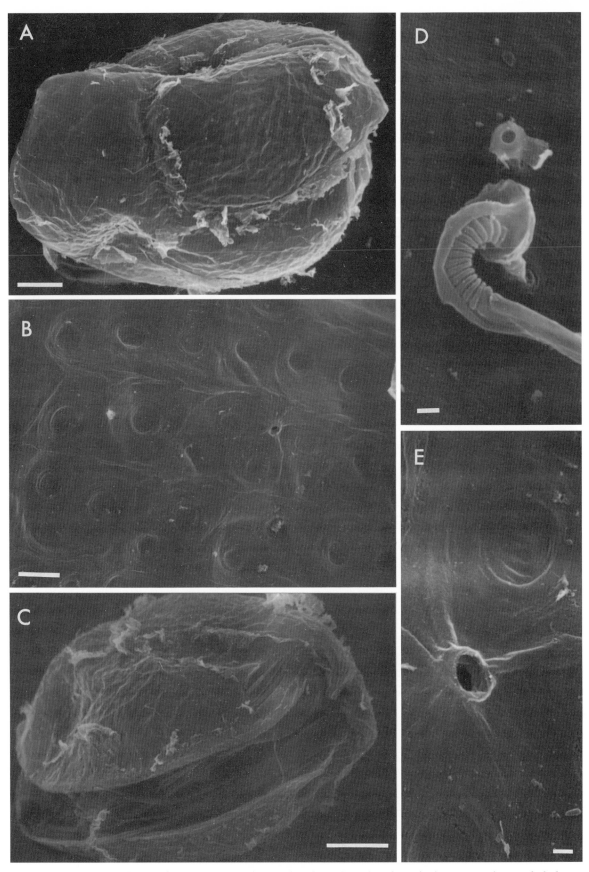

**Fig. 6.1.2A–E.** *Mikroconchoecia* cf. *acuticosta*. Female: **A** right valve; **B** lateral surface of valve; **C** ventral view of whole animal; **D** base of bristle; **E** detail of pore. *Bars* **A,C** 100 μm; **B** 10 μm; **D,E** 1 μm

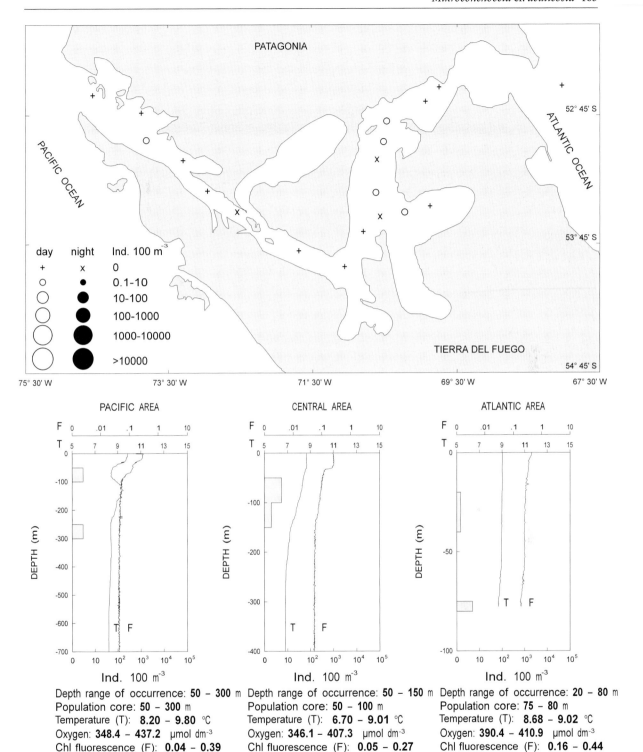

**Fig. 6.1.3.** *Mikroconchoecia* cf. *acuticosta*. Distribution in the Straits of Magellan in February–March 1991

## 6.2 *Discoconchoecia* aff. *elegans* Sars, 1865

*Conchoecia elegans* Sars, 1865: p. 117 – Müller, 1906a: p. 69, pl. 13 figs. 10, 11, 19–26 – Skogsberg, 1920: pp. 624–636, figs. 117, 118 – Angel, 1981: p. 558, fig. 194/41.
*Paraconchoecia gracilis* Claus, 1890: p. 15.
*Conchoecia quadrangularis* Aurivillius, 1898: p. 42.
*Paraconchoecia elegans* Poulsen, 1973: pp. 45–47, fig. 18.
*Conchoecia (Paraconchoecia) elegans* Deevey, 1978: pp. 51, 52, fig. 5 – Deevey, 1982: p. 147, fig. 16.
*Discoconchoecia* aff. *elegans* Martens, 1979: p. 344.

Carapace length –
    females: 1.56–1.67 mm
    males: 1.66–1.71 mm

**Female.** Adult shell elongate and rectangular; the height:length ratio for most adults about 0.43 – 0.45; rostrum rather short and slightly curved; incisure deep; dorsal margin straight below well-developed shoulders; posterior margin almost straight; postero-dorsal terminus indistinctly pointed in the right but not in the left valve; ventral margin regularly convex. Asymmetric glands in the usual places; lateral corner glands and dorsomedial glands absent; medial glands prominent, about 9–10 posteriorly and numerous along the ventral margin; about 6 elongate anterior glands below the incisure. Shell sculpture observed to be finely reticulate on many specimens. Cap of the frontal organ thickened in the middle and rounded in front. Al without a dorsal bristle; of the 5 end bristles, the 'e' bristle is slender, long and carries spinules, the 4 sensory 'pipe' bristles are short and simple. Furca with 8 claws separated from the lamella and reducing in size rearwards; no posterior unpaired bristle.

**Male.** Carapace similar to that of the female, except for the occurrence of a postero-dorsal dorsomedial gland on each valve. Cap of the frontal organ clearly jointed and produced frontally; ornament of scattered stiff ventral hairs and numerous minute follicles. The 'e' bristle of the A1 carries a disc-like expansion (principal generic character) and below this 1–2 stout spinules. A2 clasping joints dimorphic; the right one more strongly recurved than the left one. P2 larger than in the female (although bearing shorter bristles). Penis with a backwards-curving rounded tip; about 5 transverse muscle bands; seminal tube simple and broadened at its base.

### Remarks

In our specimens, the mean length of mature males is a little greater than that of mature females. This is unusual among planktonic ostracods, in which females are generally larger, as noted by Poulsen (1973). The size range of our material is more restricted than the 1–2 mm cited by Angel (1981) for adult females; but he also notes that there are several different 'forms' for this species, those from lower latitudes being smaller than those from high latitudes. For Koch (1992), who also works on Antarctic ostracods, this difference can be characterized at the subspecific level and he refers to *D. elegans gracilis* as equal to the 'southern form' of Angel (1972). Chavtur (1993) considers that a number of new species are hidden under the old species names in this genus. However, no author as yet has been able to characterize the different forms of *D. elegans* sufficiently for them to be consistently recognized by others working on different collections. In our material, ovigerous females are about 3 times as numerous as mature males in deeper water (300–200 m) and about 6 times as numerous as mature males near the surface (40–20 m). We identified all juvenile stages, except the 2 smallest, which were too small for the mesh of the nets used and thus escaped capture.

## Distribution

With the caveat (Chavtur 1993) noted above, this species appears to be completely cosmopolitan, occurring in all oceans. It ranges between about 70°N and 65°S, and has been taken at depths up to 5190 m (Deevey 1983) although, as noted by Hartmann (1985), it is most usually sampled at less than 1000 m. As with the previous species, if the deep water records are not of contaminants, they may be indices for Antarctic Bottom Water at lower latitudes.

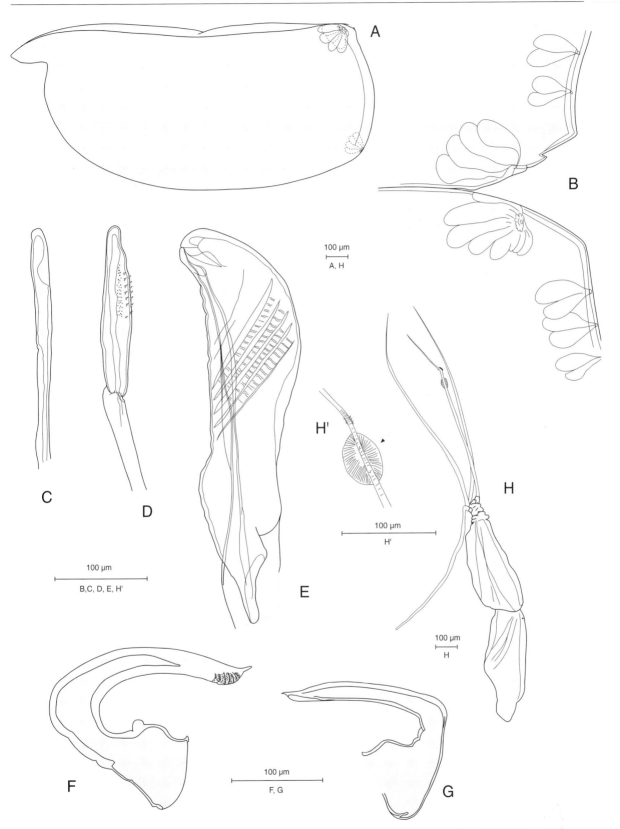

**Fig. 6.2.1A–H.** *Discoconchoecia* aff. *elegans*. Female: **A** shell of ovigerous adult; **B** posterior margin of shell; **C** frontal organ. Male: **D** frontal organ; **E** penis; **F** A2 right clasping organ; **G** A2 left clasping organ; **H** A1; **H'** suctorial organ of the 'e' bristle

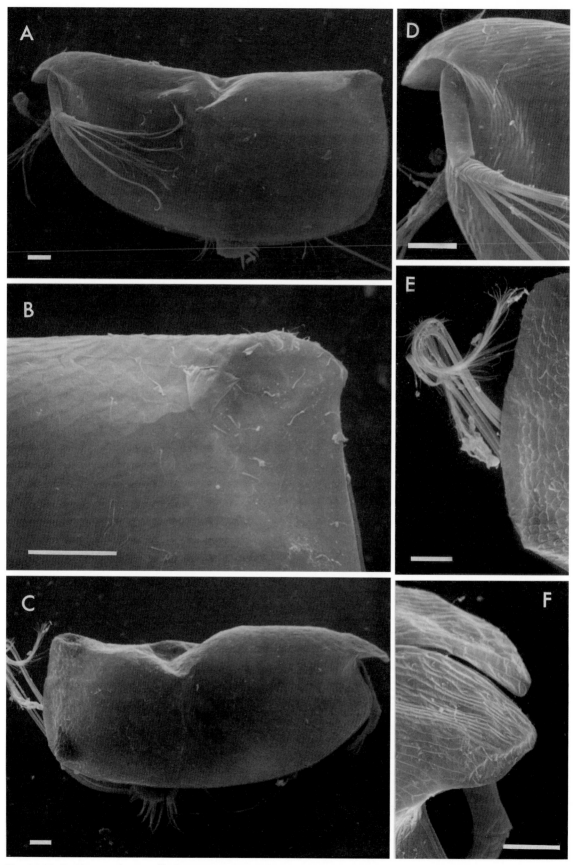

**Fig. 6.2.2A–F.** *Discoconchoecia* aff. *elegans*. Female: **A** left valve; **B** posterior caudal process; **D** detail of the rostrum. Male: **C** right valve; **E** posterior caudal process; **F** detail of the rostrum. *Bars* **A,B,C,D,E,F** 100 µm

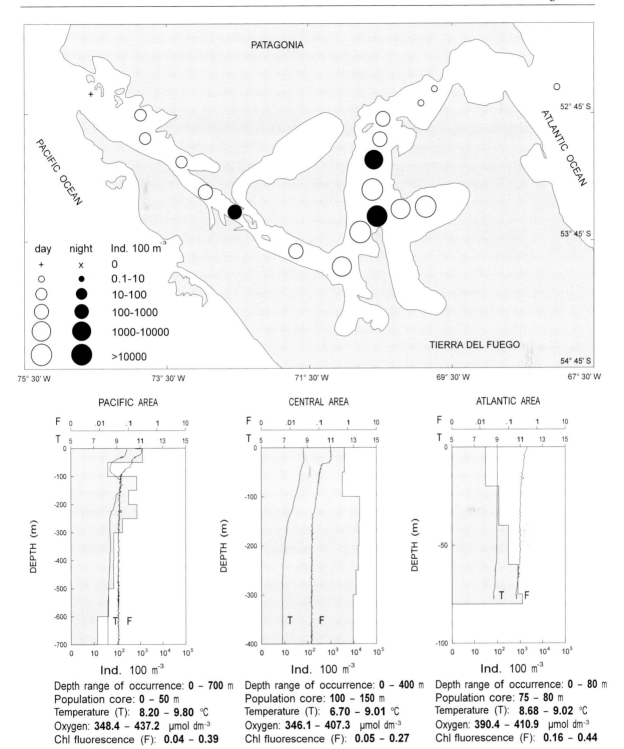

**Fig. 6.2.3.** *Discoconchoecia* aff. *elegans*. Distribution in the Straits of Magellan in February–March 1991

## 6.3 *Loricoecia loricata* Claus, 1894

*Conchoecia loricata* Claus, 1894: p. 4, pl. 3 figs. 24–30 – Angel, 1981: p. 558, fig. 194/56.
*Conchoecia loricata typica* Müller, 1906a: p. 95, pl. 22 figs. 1–9.
*Conchoecia loricata loricata* Müller, 1912: pp. 80, 81.
*Loricoecia loricata* Poulsen, 1973: pp. 145–147, figs. 72, 73.
*Conchoecia (Loricoecia) loricata* Deevey, 1978: p. 60, fig. 15d–f – Deevey, 1982: pp. 157, 158, fig. 30.

Carapace length –
    females: 2.05 mm
    males: 1.86 mm

**Female.** Carapace large and regularly rectangular; height about 1/2 the length; rostrum almost straight; incisure open and moderately deep; dorsal margin weakly concave below the shoulders; posterior margin convex, making a blunted right angle with the end of the dorsal margin; ventral margin slightly convex. Asymmetric glands at the usual sites; lateral corner glands present; medial glands numerous posteriorly. Shell sculpture consists of distinct striae which run parallel to the dorsal and posterior margins; from the incisur they curve towards the ventral margin, thus intersect that margin in the anteroventral region, giving a denticulate effect; a weak cross-striation is observable anteriorly. Cap of the long frontal organ is united with the stem and produced distally; ornament of many short and stiff hairs everywhere except dorsodistally. Al segments spinulose, with a dorsal bristle; the 5 end bristles comprise a long slender 'e' bristle and 4 short sensory 'pipe' bristles.

**Male.** Carapace similar to that of the female but smaller and with paired dorsomedial glands posteriorly. Cap of the frontal organ separated from the stem; ornament of short and rather stiff hairs, except dorsodistally, where it is smooth. A1 'e' bristle bearing a double row of 20–22 backwards-directed 'thorns' (short spines). A2 clasping joints weakly dimorphic, both carrying a small basal protuberance and right-angled at the elbow; the right clasping joint somewhat the larger and more recurved. Penis with convex margins (one seems broadly serrulate) and a flattened top; several transverse muscle bands; seminal tube broadened basally and also thickened medially.

### Remarks

In size, our specimens are within and towards the upper part of the range cited by Angel (1981), viz. females 1.8–2.1 mm, males 1.6–1.9 mm. The rectangular shape and convex posterior margin, also the reticulation pattern of the carapace, are other features which allow its ready recognition.

### Distribution

Müller (1912) notes that this species occurs widely in the Atlantic, Pacific and Indian Oceans and the Mediterranean. For Angel (1981) it is a mesopelagic and transoceanic taxon, ranging in the Atlantic from 60°N to 47°S. Poulsen (1973, fig. 73) shows many records around Indonesia. Deevey (1978, 1982) adds that it ranges in the Pacific from 11°N to 52°S, and from equatorial regions to 50°S in the Indian Ocean. In the Magellan Straits samples, it is confined to the western part of the sampling area and is never common. The depth range given by Hartmann (1985) is 0–2500 m. However, our records, which are the most southerly yet for this species, are from shallow to intermediate depths.

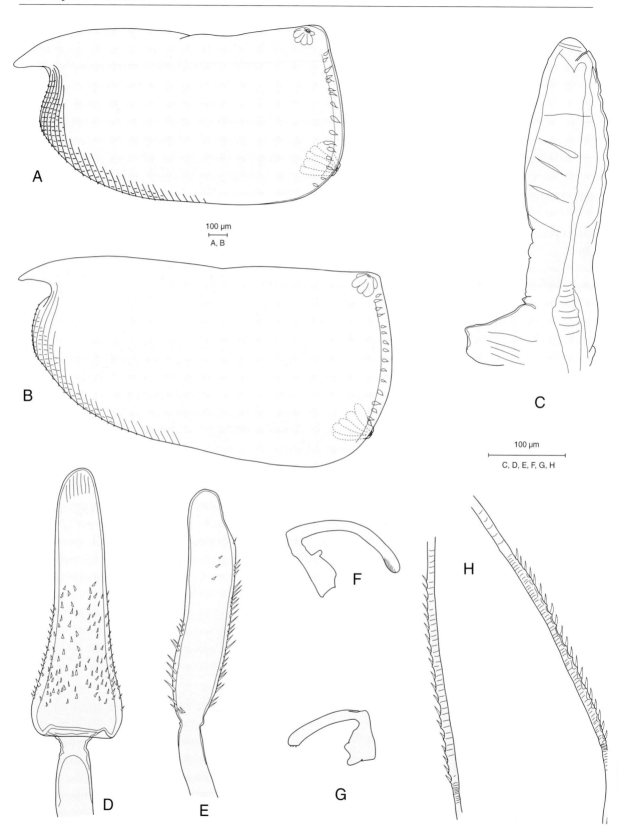

**Fig. 6.3.1A–H.** *Loricoecia loricata.* **A** shell of adult male; **B** shell of adult female; **C** penis; **D** male frontal organ; **E** female frontal organ; **F** male A2 left clasping organ; **G** male A2 right clasping organ; **H** male A1 (antennule) detail of 'e' bristle

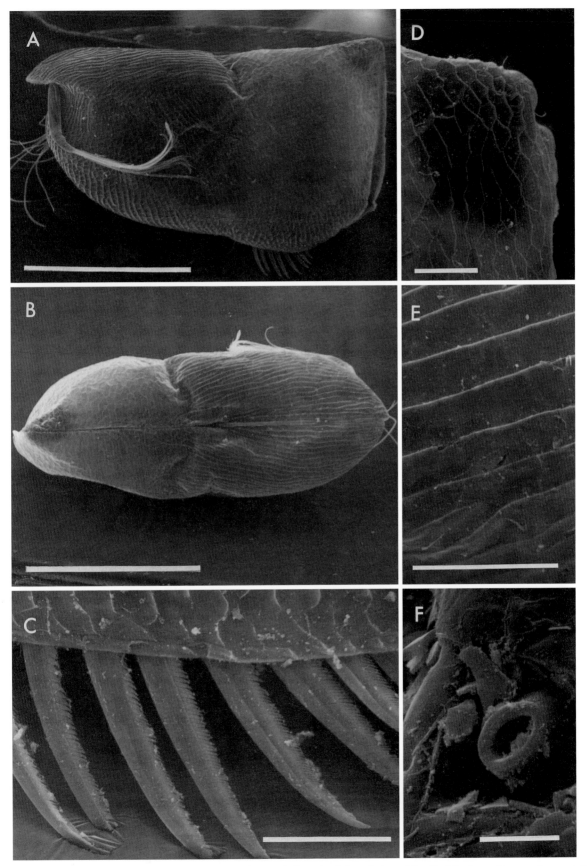

**Fig. 6.3.2A–F.** *Loricoecia loricata*. Female: **A** left valve; **B** whole animal dorsal view; **C** detail of the furca; **D** posterior caudal process; **E** postero-dorsal valve striation; **F** detail of pore. *Bars* **A,B** 1 mm; **C,D,E** 100 μm; **F** 10 μm

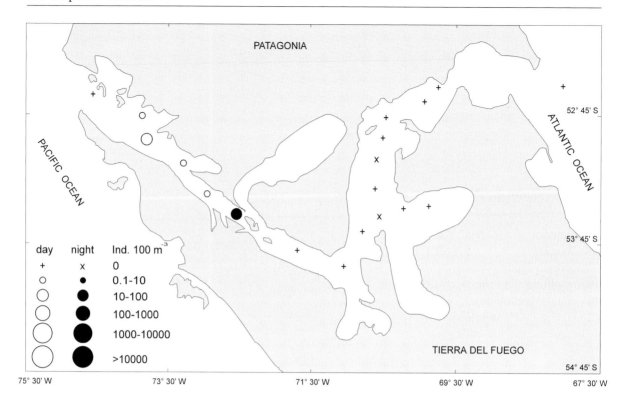

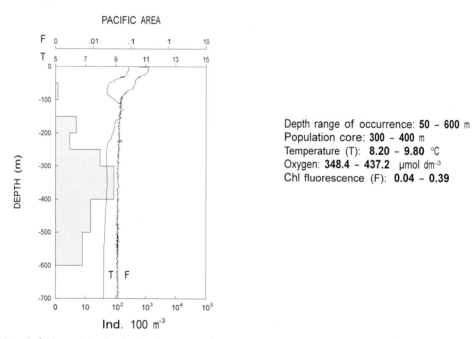

Depth range of occurrence: **50 – 600** m
Population core: **300 – 400** m
Temperature (T): **8.20 – 9.80** °C
Oxygen: **348.4 – 437.2** µmol dm⁻³
Chl fluorescence (F): **0.04 – 0.39**

**Fig. 6.3.3.** *Loricoecia loricata.* Distribution in the Straits of Magellan in February–March 1991

## 6.4 *Paramollicia rhynchena* Müller, 1906

*Conchoecia rhynchena* Müller, 1906a: p. 113, pl. 26 figs. 17–25 – Angel, 1981: p. 559, fig. 194/79 – Deevey, 1982: p. 160.
*Paramollicia rhynchena* Poulsen, 1973: pp. 176–179, figs. 91, 92.
*Conchoecia (Paramollicia) rhynchena* Deevey, 1978: p. 64.

Carapace length –
    females: 2.48 mm
    males: 2.35 mm

**Female.** Carapace large and subrectangular; height slightly less than 1/2 the length, greatest height posterior; rostrum curved; incisure rather broad and shallow; dorsal margin straight below the shoulders; posterior margin weakly convex; ventral margin nearly straight and sloping backwards slightly towards the anterior. Shell sculpture consists of several very distinct ventral striae, which run parallel to the margin from the incisur almost to the posteroventral edge, with fine reticules between them; remainder of shell smooth or weakly ornamented. Asymmetric glands at the usual locations; lateral corner and dorsomedial glands present (note that both sexes have dorsomedial glands in this genus); medial glands occur along the inner posterior margin. Cap of the frontal organ jointed onto the stem and produced distally where it is subacuminate; ornamented with numerous stiff hairs, except distally. Al with a long dorsal bristle and 5 end bristles; one of these is long and weakly spinulose, the 4 others are short, sensory 'pipe' bristles.

**Male.** Carapace similar to that of the female, but smaller. Cap of the frontal organ jointed from the stem and expanded basally but narrowed distally; ornamented with numerous stiff hairs, except distally. Al with 5 end bristles; of these, the chief ('e') bristle is adorned with a double row of numerous small blade-like backwards-directed spines, the 'b' bristle is about as long and carries a short thickened extension, the 'd' bristle is spinulose, the 'a' bristle is long and reflexed, and the 'c' bristle is a short 'pipe' bristle. The A2 clasping joints are dimorphic; the right is larger, right-angled at the elbow then strongly recurved, basally it has 2 small protuberances; the smaller left clasping joint is bent backwards at the elbow and only slightly recurved. Penis large, with parallel sides and an obliquely truncate tip; about 7 transverse muscle bands; seminal tube simple and rather wide.

### Remarks

This is a striking and large species, easily recognized by the prominent striations parallel to the ventral margin and the rather elongate lateral profile. In size, our specimens are about in the middle of the range cited by Angel (1981), viz. females 2.3–2.9 mm, males 2.2–2.6 mm.

### Distribution

Angel (1981) regards this by and large as a mesopelagic and transoceanic species. Poulsen (1973, fig. 92) records it from the eastern Atlantic, northern Indian and southwestern Pacific Oceans. Deevey (1982) gives the latitudinal range as 60°N to 46°S in the Atlantic, and to 49°S in the Pacific. Like *Loricoecia loricata*, in the Magellan Straits this species is confined to the western part of the sampling area, where it is fairly common; this is the most southerly record for the species. Hartmann (1985) gives its depth range as 250–3000 m, with most records between 1500 and 3000 m. Our specimens were taken at intermediate depths.

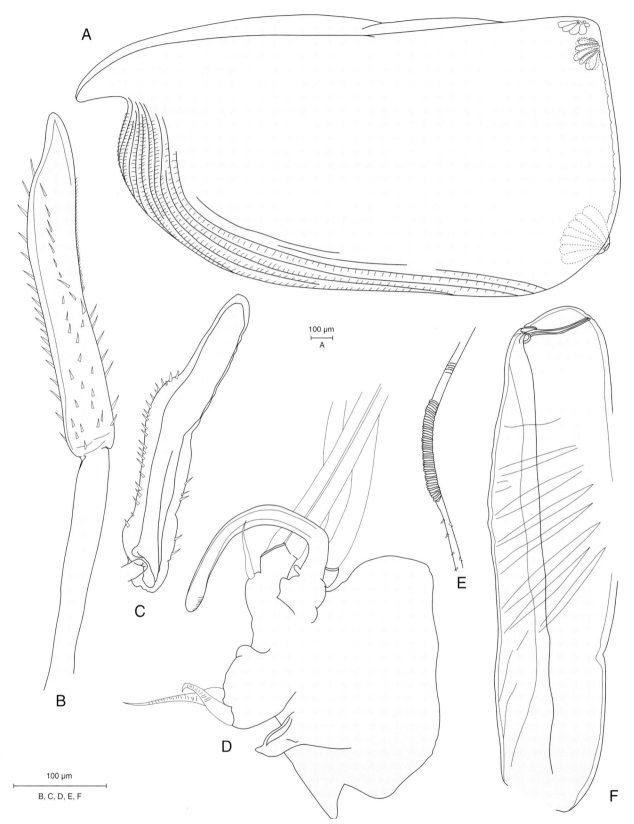

**Fig. 6.4.1A–F.** *Paramollicia rhynchena*. Female: **A** shell of ovigerous adult; **B** frontal organ. Male: **C** frontal organ; **D** A2 right clasping organ; **E** A1 (antennule) detail of 'e' bristle; **F** penis

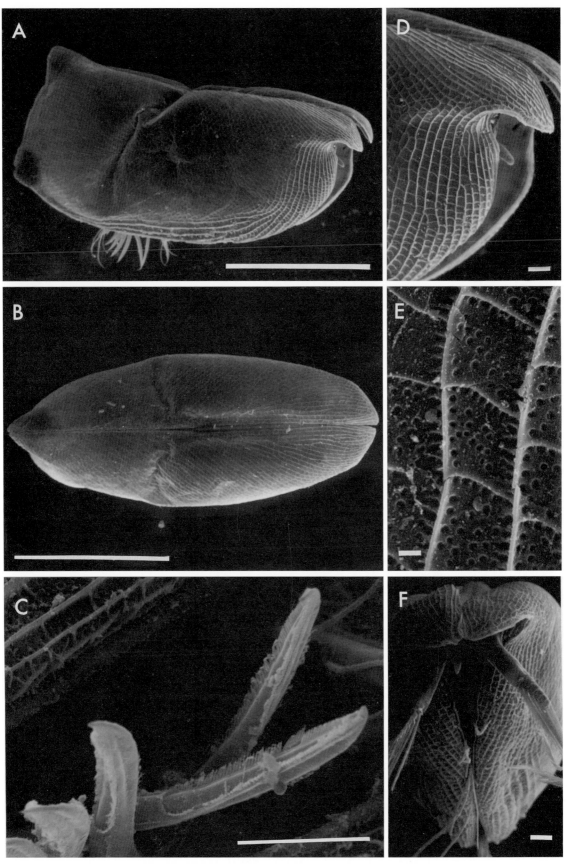

**Fig. 6.4.2A–F.** *Paramollicia rhynchena.* Female: **A** right valve; **B** whole animal dorsal view; **C** detail of the furca; **D** detail of the rostrum; **E** detail of valve reticulation, **F** whole animal frontal view. *Bars* **A,B** 1 mm; **C,D,F** 100 μm; **E** 10 μm

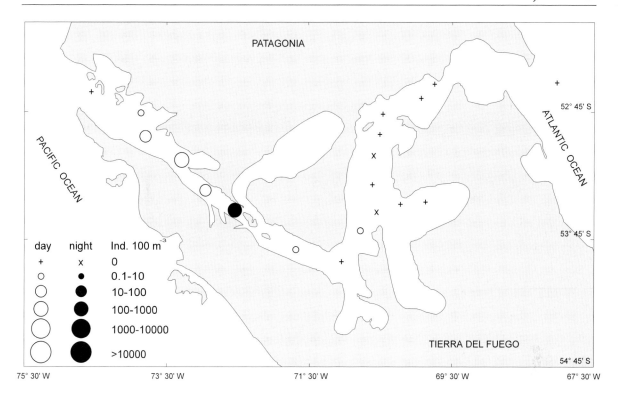

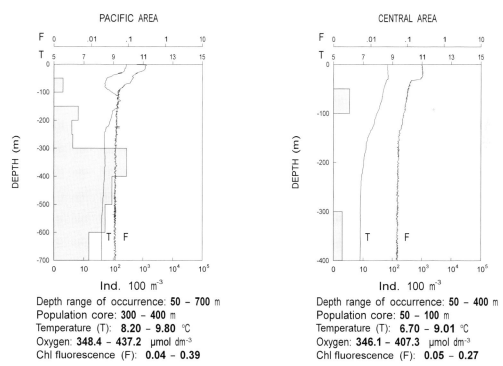

Depth range of occurrence: **50 – 700** m
Population core: **300 – 400** m
Temperature (T): **8.20 – 9.80** °C
Oxygen: **348.4 – 437.2** μmol dm⁻³
Chl fluorescence (F): **0.04 – 0.39**

Depth range of occurrence: **50 – 400** m
Population core: **50 – 100** m
Temperature (T): **6.70 – 9.01** °C
Oxygen: **346.1 – 407.3** μmol dm⁻³
Chl fluorescence (F): **0.05 – 0.27**

**Fig. 6.4.3.** *Paramollicia rhynchena.* Distribution in the Straits of Magellan in February–March 1991

## 6.5 *Metaconchoecia australis* Gooday, 1981 *nov. comb.*

*Conchoecia rotundata* Müller, 1906a: pp. 83, 84, pl. 17 figs. 23-34 [*partim* ] – Müller, 1908: pp. 69, 70 – Skogsberg, 1920: pp. 649–658, figs. 122, 123 [*partim* ].
*Conchoecia skogsbergi* Iles, 1953 – Angel, 1981: p. 564, fig. 194/88 [*partim* ?].
*Conchoecia australis* Gooday, 1981: pp. 154, 155, figs. 37, 38.

Carapace length –
    females: 0.98–1.06 mm
    males: 0.90–1.00 mm

**Female.** Carapace small, subovate; height a little more than 1/2 the length; rostrum short, curved; incisure shallow; dorsal margin straight below the shoulders; posterior margin broadly rounded continuing smoothly into the convexly curved ventral margin. Right asymmetric gland posterodorsal, near the fusion of the valves; left asymmetric gland on a small anterodorsal node, near the base of the rostrum; lateral corner glands absent. Shell smooth. Cap of the short frontal organ united with the stem but thickened and finger-shaped, tip subacuminate, adorned dorsally and ventrally with short hairs except near the tip. A1 short, without a dorsal bristle; of the 5 end bristles, the 'e' bristle is relatively long and smooth, the 'a'–'d' bristles are shorter, sensory 'pipe' bristles.

**Male.** Carapace similar to that of the female but smaller. Cap of the short frontal organ separated from the stem by a joint, rather stout and bent in the middle, tip rounded; ornamented ventrally with short hairs and carrying 6–8 spinules prox-imodorsally. A1 'e' bristle bearing a double row of 12–15 thorn-like spines, 'b' and 'd' bristles almost as long and weakly spinulose, 'c' bristle a short sensory 'pipe' bristle, 'a' bristle reflexed. Clasping joints of the A2 dimorphic; right one larger and strongly recurved, left one smaller. Penis small, parallel-sided, with a flattened tip; 3–4 transverse muscle bands; seminal tube simple.

### Remarks

In our material, the size of females and males remains within the range given above, but Angel (1981) cites a length of 1.06 mm. We originally followed Angel in determining the species as *M. skogsbergi* (Benassi et al. 1994) but subsequently obtained a copy of Gooday (1981) through the kind assistance of Drs. Angel and Gooday. Using Gooday's approach, this species clearly is *australis*. The carapace outline matches his drawings in fig. 37 (Gooday 1981). Further, the left asymmetric gland is located at almost 13 % of the length behind the tip of the rostrum, and the carapace height is more than 50 % of the length. The generally small size of *M. australis* ensures that juveniles are regularly missed by nets with a mesh coarser than 250.

### Distribution

For Gooday (1981), *M. australis* occurs in 5 stations situated between 39°S and 50°S in the SW and SE Atlantic Ocean; as well as in one *Gauss* station in the SW Indian Ocean. The depth range of the material studied by Gooday is 250–750 m, with most occurrences between 500 m and 750 m. In the Magellan Straits, it is confined to the western part of the sampling area at depths usually less than 500 m.

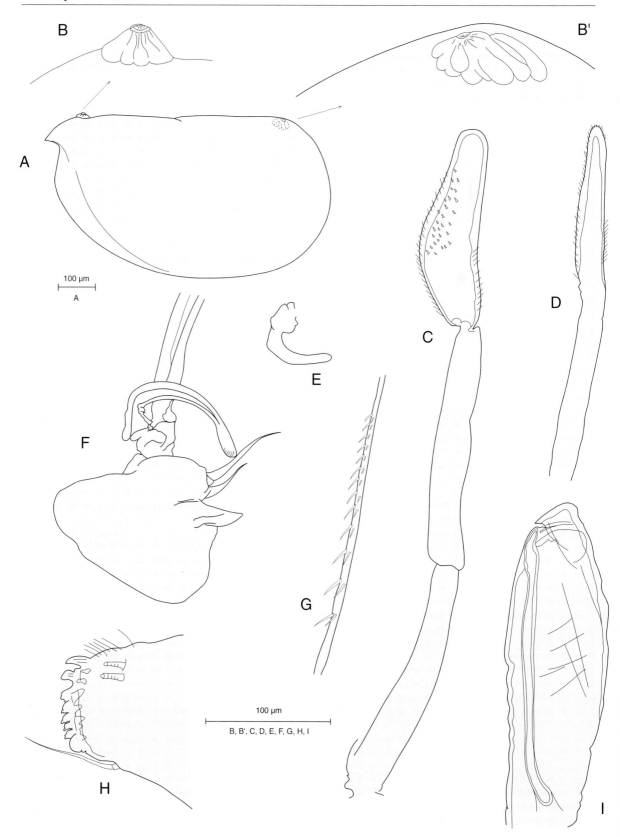

**Fig. 6.5.1A–I.** *Metaconchoecia australis.* **A** shell of adult female; **B** shell of adult female, detail of anterior glands; **B′** shell of adult female, left posterodorsal valve; **C** male frontal organ; **D** female frontal organ; **E** male A2 left clasping organ; **F** male A2 right clasping organ; **G** male A1 (antennule) detail of 'e' bristle armature; **H** male mandible; **I** penis

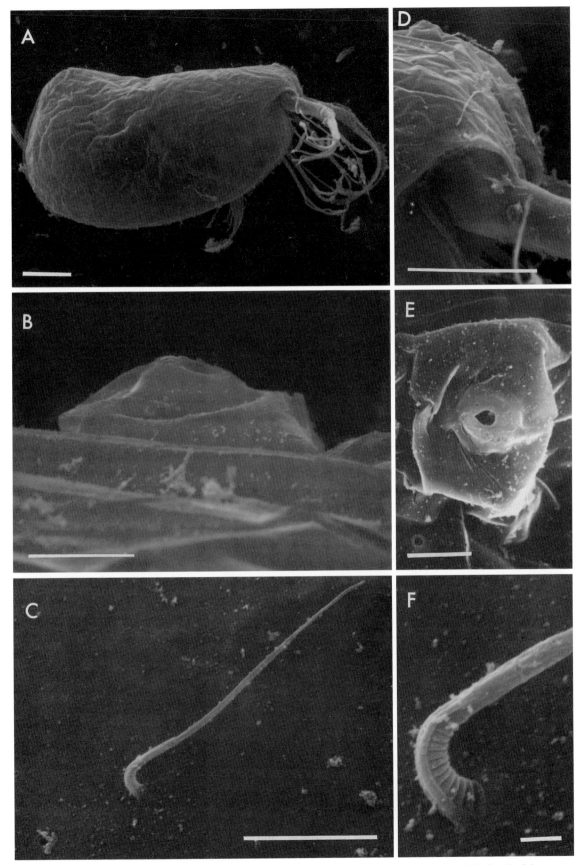

**Fig. 6.5.2A–F.** *Metaconchoecia australis*. Male: A right valve; **B,E** detail of the asymmetric gland; **D** detail of the rostrum; **C,F** detail of the bristle. *Bars* **A,D** 100 μm; **B,C,E** 10 μm; **F** 1 μm

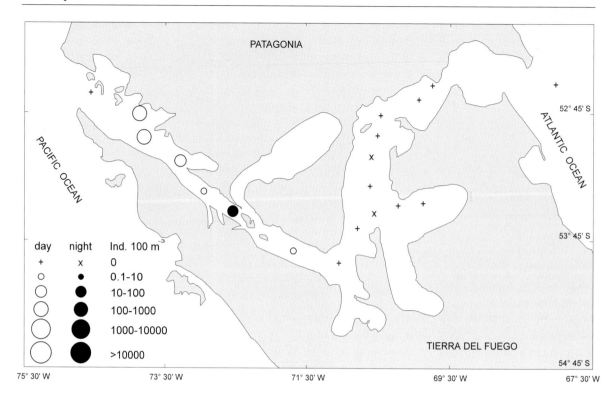

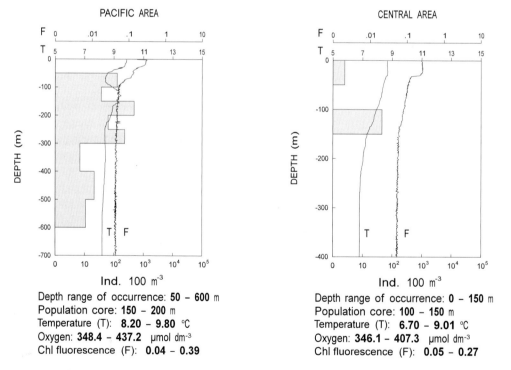

Depth range of occurrence: **50 – 600** m
Population core: **150 – 200** m
Temperature (T): **8.20 – 9.80** °C
Oxygen: **348.4 – 437.2**  μmol dm⁻³
Chl fluorescence (F): **0.04 – 0.39**

Depth range of occurrence: **0 – 150** m
Population core: **100 – 150** m
Temperature (T): **6.70 – 9.01** °C
Oxygen: **346.1 – 407.3**  μmol dm⁻³
Chl fluorescence (F): **0.05 – 0.27**

**Fig. 6.5.3.** *Metaconchoecia australis.* Distribution in the Straits of Magellan in February–March 1991

## 6.6 *Conchoecilla* cf. *chuni* Müller, 1906

*Conchoecia chuni* Müller, 1906a: pp. 124, 125, pl. 31 figs. 16–28 – Müller, 1912: pp. 93, 94, fig. 25 – Angel, 1981: p. 557, fig. 194/27.
*Conchoecilla chuni* Poulsen, 1973: pp. 221–223, figs. 112, 113.
*Conchoecia (Conchoecilla) chuni* Deevey, 1978: pp. 68, 69, fig. 22 – Deevey, 1982: p. 165, fig. 38.

Carapace length –
   females: 2.15 mm
   males: 1.40–1.55 mm (Müller, 1906a)

**Female.** Carapace large and boat-shaped; rostrum long and curved, asymmetrical, shorter in the right than in the left valve, tip acuminate; incisure rather narrow and deep; dorsal margin straight or weakly concave below the gently convex shoulders; right posterodorsal edge denticulate, tip pointed, left edge shorter with a blunted tip; posterior margin convex making a smooth curve with the ventral margin. Right asymmetrical gland located anteriorly below the incisure; left asymmetric gland sited on the posterior margin, dorsalwards of its smooth junction with the ventral margin; additionally there are 5 large lateral glands on the posterior margin; medial glands, however, are indistinct. Shell of our specimen smooth. Cap of the frontal organ united with the stem and not much thicker, tip rounded; ornamented ventrally with short hairs. Al with a dorsal bristle; end bristles comprising a long spinulose chief ('e') bristle and 4 shorter sensory 'pipe' bristles.

**Male.** Carapace much shorter than in the female, not as produced posteriorly, and more distinctly reticulate but otherwise similar. Cap of the frontal organ separated from the stem by a joint; finger-like, with a rounded tip, and smooth. Al 'e' bristle carrying a double row of about 11–13 longish, slender, backwards-directed spines and distally of these a small clump of hairs, 'b' and 'd' bristles nearly as long, 'c' bristle a short 'pipe' bristle, 'a' bristle short and strongly reflexed. A2 clasping joints dimorphic; right joint larger with a single basal protuberance and strongly recurved, left joint bent back at the elbow; both joints have spinules at their tips. Penis almost spatulate; broader distally, where it has a flat top, than at its base; 6–7 weak transverse muscle bands; seminal tube simple and slender (Müller 1906a).

### Remarks

The genus *Conchoecilla* is easily recognized by a relatively elongate carapace which is pointed posteriorly in both valves (the points unequal in length), but more particularly by the position of the right asymmetric gland which occurs anteriorly, below the incisure, in the right valve rather than at or near the postero-ventral corner of this valve as in most conchoeciines. The unusual shape, the positions of the two asymmetric glands and the distinctive lateral glands enable this genus to be readily distinguished. As a rule, the types and locations of asymmetric glands and the presence/absence of lateral corner glands, dorso-medial glands, and medial glands are all very useful characters in defining the various conchoeciine genera.

### Distribution

We have a single ovigerous female in sample 3 at Station 10 (depth 400–500 m). Iles (1953) records a single female from the Benguela Current. Deevey notes that it is a Southern Hemisphere species and summarizes its range of occurrence as 26° to 55°S in the Atlantic, 37° to 64°S in the Pacific and 2° to 44°S in the Indian Ocean. It is never abundant, and small numbers are found throughout the water column. Angel (1981) regards it as a mesopelagic species. Hartmann (1985) gives its depth range as 0–3750 m, noting that it is most usually taken between 30–600 m.

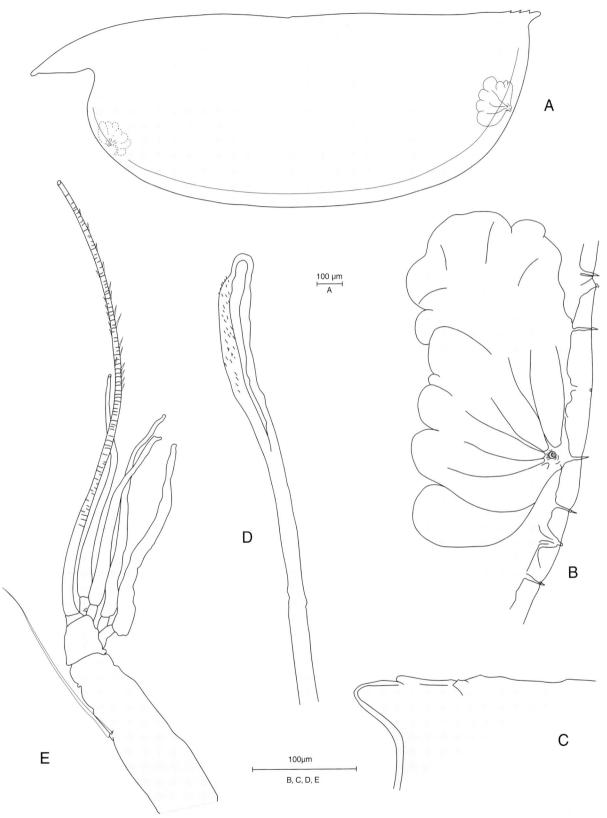

**Fig. 6.6.1A–E.** *Conchoecilla* cf. *chuni*. Female: **A** shell of ovigerous adult; **B** glands of posterior left valve; **C** detail of posterior margin; **D** frontal organ; **E** A2

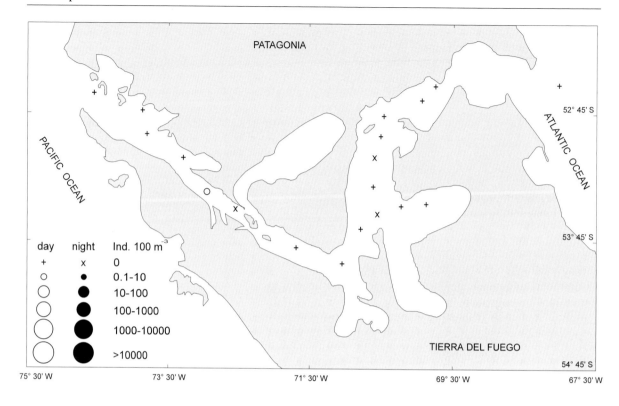

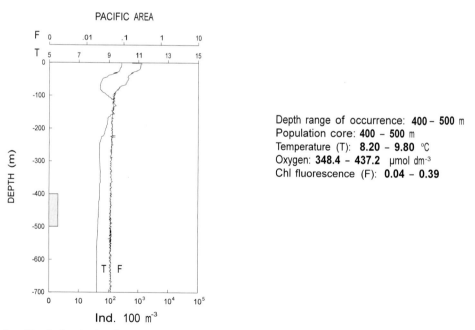

Depth range of occurrence: **400 – 500** m
Population core: **400 – 500** m
Temperature (T): **8.20 – 9.80** °C
Oxygen: **348.4 – 437.2** µmol dm⁻³
Chl fluorescence (F): **0.04 – 0.39**

**Fig. 6.6.2.** *Conchoecilla* cf. *chuni*. Distribution in the Straits of Magellan in February–March 1991

## 6.7 Pseudoconchoecia serrulata Claus, 1874

*Conchoecia serrulata* Claus, 1874a: p. 176 – Claus, 1874b: pl. 1 figs. 2–7, 9, 10, pl. 2 figs. 12, 13, 17, 19 – Müller, 1906a: p. 97, pl. 22 fig. 24, pl. 23 figs. 1–13 – Skogsberg, 1920: pp. 681–688, fig. 130 – Ramirez and Moguilevsky, 1971: p. 659, pl. 16 figs. 1–5, pl. 17 figs. 1–5.
*Halocypris atlantica* Brady, 1880: p. 164, pl. 40 figs. 1–15, pl. 41 figs. 11, 12.
*Conchoecia serrulata serrulata* Müller, 1912: pp. 81, 82 – Angel, 1981: p. 560, fig. 194/87.
*Pseudoconchoecia serrulata* Claus, 1891: p. 72, pl. 19 figs. 1–14, pl. 23 figs. 1–13 – Poulsen, 1973: pp. 148, 149, fig. 74.
*Conchoecia (Pseudoconchoecia) serrulata* Deevey, 1978: p. 61, fig. 16 – Deevey, 1982: p. 158, fig. 31.

Carapace length –
  females: 1.47 mm
  males: 1.25 mm

**Female.** Carapace of medium size, rectangular; height:length ratio about 0 : 60; rostrum short, curved; incisure broad; dorsal margin straight below the well-developed shoulders; posterodorsal edges of both valves slightly produced but rounded; posterior margin convex, curving smoothly into the ventral margin, which is weakly convex or straight. Asymmetric glands at the usual sites; lateral corner glands present. Shell sculpture distinct; consisting of numerous longitudinal striations tending to converge towards the rostral region, thus producing a serrulate margin below the incisure. Cap of the frontal organ united with the stem, thickened near its base and smooth. A1 with a long spinulose dorsal bristle; end bristles comprising a very long spinulose 'e' bristle and 4 short sensory 'pipe' bristles. Mandible basale relatively short (generic character).

**Male.** Carapace shorter than that of the female but otherwise similar, except that paired dorsomedial glands are present. Cap of the frontal organ separated by a joint from the stem, finger-shaped and smooth. A1 'e' bristle adorned with a single row of 20–22 broad-based spinules, the 'b' and 'd' bristles almost as long and smooth, the 'c' bristle is a short sensory 'pipe' bristle, the 'a' bristle is short and reflexed. A2 clasping joints strongly dimorphic; right joint with 2 basal protuberances, and 2 right-angled bends; left joint much smaller, also with 2 basal protuberances but recurved distally. Mandible basale relatively short, as in the female. Penis with weakly convex sides and a flattened top; 6–7 transverse muscle bands; seminal tube simple, slightly thicker medially.

### Remarks

This is a distinctive species, easily recognized by its rather squat lateral profile (higher with respect to length than in most species of this size), well-developed shoulders on each valve, striate ornament, and the serrulate valve margin below the incisure. Poulsen (1973) regards the unusually short mandibular basale as a further distinctive character.

### Distribution

Angel (1981) considers this to be an upper mesopelagic species found in all Southern Hemisphere oceans. Deevey (1982) summarizes its distribution as from 10° to 58°S in the Atlantic, 37° to 68°S in the Pacific and 36° to 59°S in the Indian Ocean. Poulsen (1973) adds a single record from Indonesian seas. It is a common species in Magellan Straits, from about 200 m to the surface; and we recorded it previously from the second Italian Antarctic expedition in considerable numbers at St. 1 and 2 (Benassi et al. 1992). Deevey (1983) gives the depth range as 0–5190 m, but earlier Deevey (1982) noted that it was most abundant in the upper few hundred metres, between the Subtropical and Antarctic Convergences.

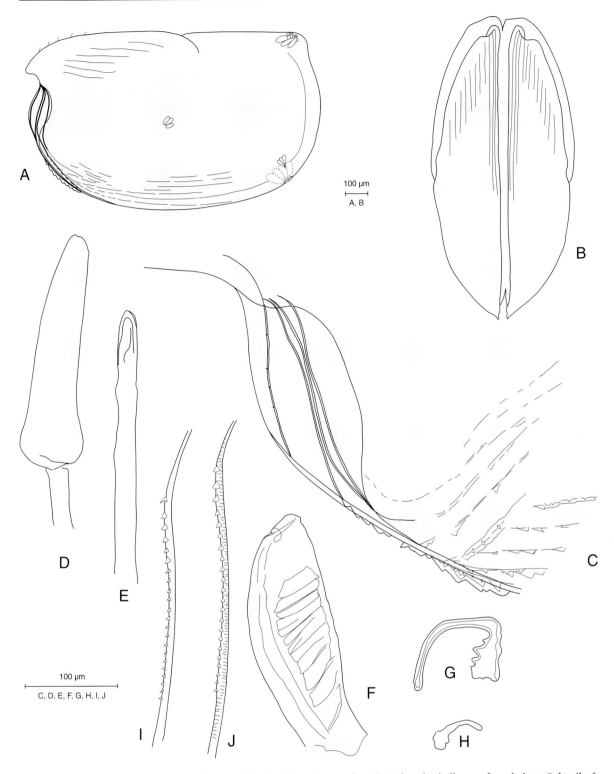

**Fig. 6.7.1A–J.** *Pseudoconchoecia serrulata.* **A** shell of adult ovigerous female; **B** female shell, seen from below; **C** detail of ventral margin of female shell; **D** male frontal organ; **E** female frontal organ; **F** penis; **G** male A2 right clasping organ; **H** male A2 left clasping organ; **I,J** male A1 (antennule) detail of 'e' bristle armature

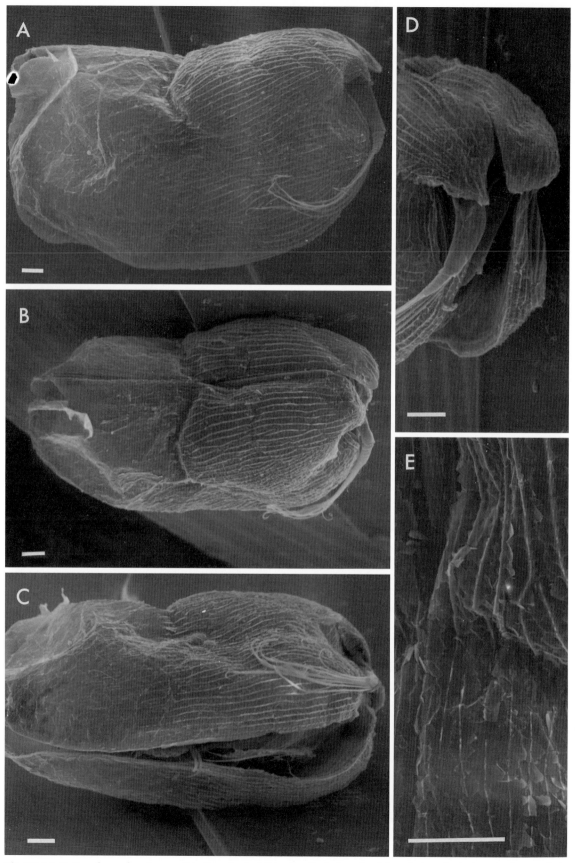

**Fig. 6.7.2A–E.** *Pseudoconchoecia serrulata.* Female: **A** right valve; **B** whole animal dorsal view; **C** whole animal ventral view; **D** whole animal frontal view; **E** detail of valve striation. *Bars* **A,B,C,D,E** 100 μm

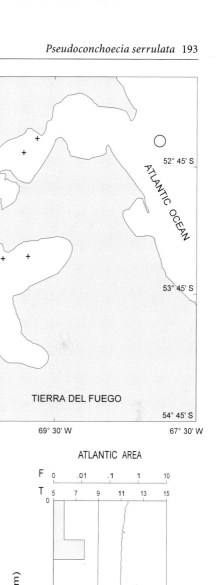

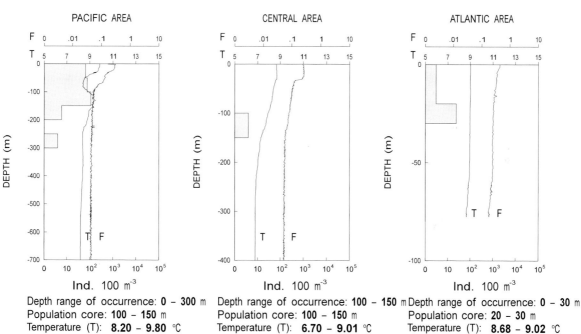

PACIFIC AREA

Depth range of occurrence: **0 – 300** m
Population core: **100 – 150** m
Temperature (T): **8.20 – 9.80** °C
Oxygen: **348.4 – 437.2** µmol dm⁻³
Chl fluorescence (F): **0.04 – 0.39**

CENTRAL AREA

Depth range of occurrence: **100 – 150** m
Population core: **100 – 150** m
Temperature (T): **6.70 – 9.01** °C
Oxygen: **346.1 – 407.3** µmol dm⁻³
Chl fluorescence (F): **0.05 – 0.27**

ATLANTIC AREA

Depth range of occurrence: **0 – 30** m
Population core: **20 – 30** m
Temperature (T): **8.68 – 9.02** °C
Oxygen: **390.4 – 410.9** µmol dm⁻³
Chl fluorescence (F): **0.16 – 0.44**

**Fig. 6.7.3.** *Pseudoconchoecia serrulata.* Distribution in the Straits of Magellan in February–March 1991

## 6.8 *Obtusoecia antarctica* Müller, 1906

*Conchoecia obtusata* Sars, 1866 var. *antarctica* Müller, 1906a: pp. 77, 78, pl. 16 figs. 10–23 – Skogsberg, 1920: pp. 647, 648 – Ramirez and Moguilevsky, 1971: p. 656, pl. 11 figs. 1–3 – Angel, 1981: p. 559, fig. 194/68.
*Spinoecia obtusata* var. *antarctica?* Poulsen, 1973: p. 123.
*Conchoecia (Spinoecia) obtusata* var. *antarctica* Deevey, 1978: p. 53, fig. 8.
*Obtusoecia antarctica* Martens, 1979: p. 332, fig. 14a-e – Koch, 1992: pp. 79–82, fig. 20.
*Conchoecia (Obtusoecia) antarctica* Deevey, 1982: p. 148, fig. 18.

Carapace length –
    females: 1.60 mm
    males: 1.37 mm

**Female.** Carapace of medium size and subovate; height about 1/2 the length; rostrum curved; incisure deep; dorsal margin straight below a rather long shoulder; posterior margin rounded, proceeding smoothly into the nearly straight ventral margin, which then curves smoothly towards the incisure. Both asymmetric glands in the usual places; lateral corner glands absent; a few long anterior glands below the incisure. Shell smooth, inner list distinct. Cap of the frontal organ united with the stem, tip pointed; ornamented with numerous short hairs. A1 with a dorsal bristle; 5 end bristles, consisting of a long smooth 'e' bristle and 4 short 'pipe' bristles.

**Male.** Carapace similar to that of the female but smaller. Cap of the frontal organ separated from its stem by a joint, finger-shaped, and densely hirsute. A1 'e' bristle adorned with numerous backwards-directed spines, distally these occur in a double row but more proximally there is only one row and they become smaller; 'b' and 'd' bristles not as long and apparently smooth, 'c' bristle a short sensory 'pipe' bristle, 'a' bristle short and reflexed. A2 clasping joints strikingly dimorphic (generic character); the right joint is extraordinarily large, with a right-angled bend at the elbow and then an obtuse-angled bend; left joint rudimentary. Penis relatively short and stout, with convex sides and a stepped top; numerous transverse muscle bands; seminal tube simple but not straight, broadened at its base.

### Remarks

Martens (1979) divided the species assigned by Poulsen (1973) to *Spinoecia* into 2 natural units to which he gave the new generic names *Obtusoecia* and *Porroecia,* thus clearing up a potential taxonomic problem. *Spinoecia* (for which the concept was too broad) then lapsed as an invalid taxon because no type species had been designated.

### Distribution

Angel (1981) regards this as an upper mesopelagic to epipelagic species in austral latitudes and all southern oceans. Deevey (1983) gives the latitudinal range as 36° to 68°S. Poulsen's questionable record is based on a single female in the *Dana* collections from south of the Fiji Islands at around 28°S. Benassi et al. (1994b) record earlier unpublished data of McKenzie citing its presence off southwestern Australia in samples collected by SS *Daimantina* during the International Indian Ocean Expedition of the 1950–1960s. We have identified this species only from the western and eastern entrances of Magellan Straits, at St. 5 and 26 respectively. Some individuals were also taken in 2 samples at St. 1 during the second Italian Antarctic Expedition (Benassi et al. 1992).

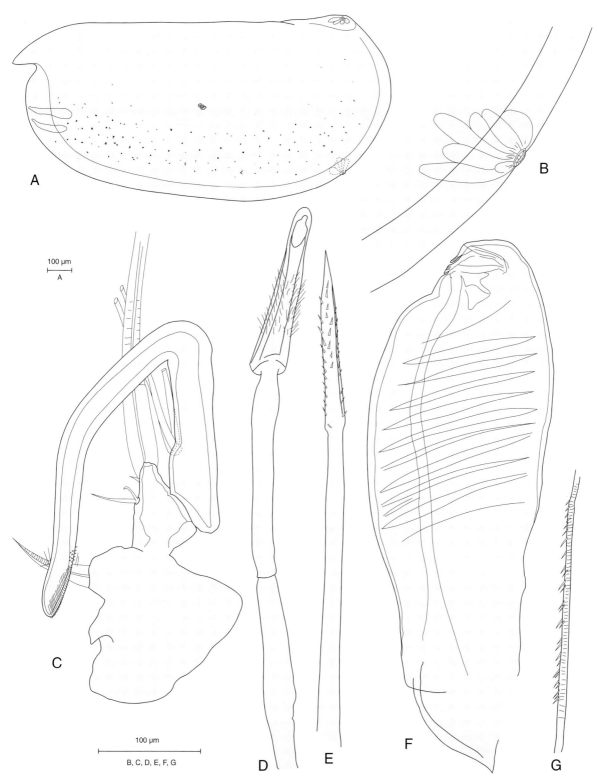

**Fig. 6.8.1A–G.** *Obtusoecia antarctica*. **A** shell of adult ovigerous female; **B** right valve female glands; **C** male A2 right clasping organ; **D** male prehensile organ; **E** female prehensile organ; **F** penis; **G** male A1 detail of 'e' bristle

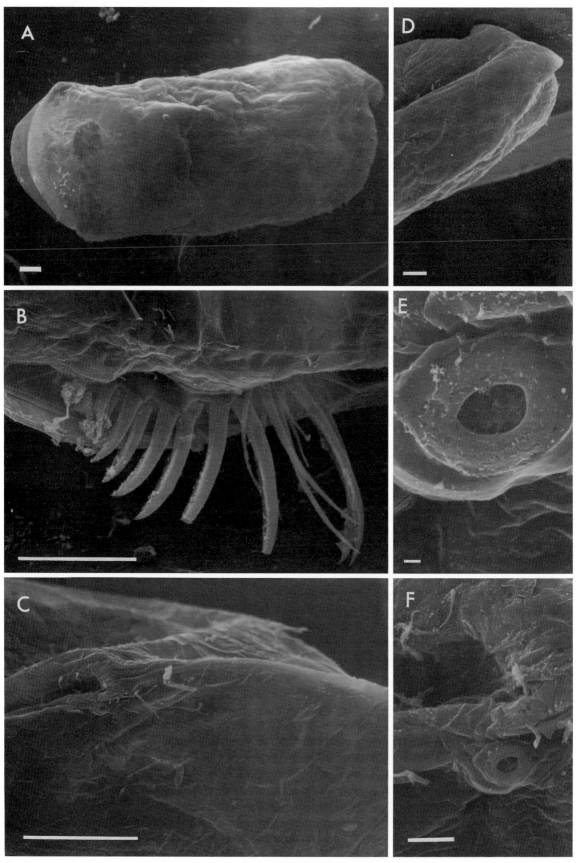

**Fig. 6.8.2A–F.** *Obtusoecia antarctica.* Female: **A** right valve; **B** detail of the furca; **C,E,F** detail of asymmetric gland in the right valve; **D** detail of the rostrum. *Bars* **A,B,C,D** 100 μm; **E** 1 μm; **F** 10 μm

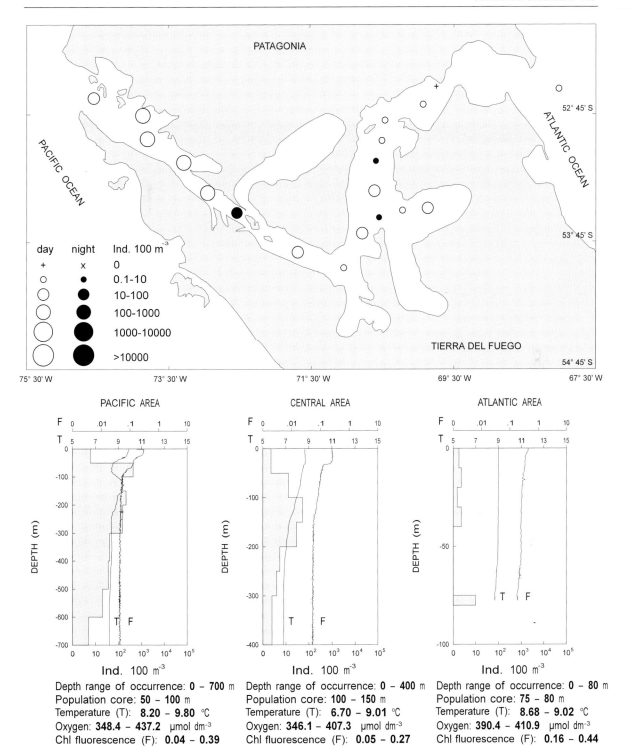

**Fig. 6.8.3.** *Obtusoecia antarctica.* Distribution in the Straits of Magellan in February–March 1991

## Order Myodocopida
Planktonic Myodocopida

### Family Cypridinidae

### 6.9  *Macrocypridina poulseni* Martens, 1979

*Cypridina castanea* Brady, 1907: p. 88, pl. 16 figs. 1–4 – Müller, 1906a: pp. 130, 131, pl. 5 figs. 1, 2, pl. 33 figs. 11–16, pl. 34 figs. 10–13 [*nec poulseni*].
*Cypridina (Macrocypridina) castanea* Skogsberg, 1920: pp. 281–287, figs. 47–51 [*nec poulseni*].
*Macrocypridina castanea* (Brady, 1897) var. *rotunda* Poulsen, 1962: pp. 136–145, figs. 70–74.
*Macrocypridina castanea* Angel, 1981: p. 555, fig. 194/3 [*partim ?*].
*Macrocypridina poulseni* Martens, 1979: pp. 307, 308, fig. 2.

Carapace length –
   females: 5.50–7.60 mm (Poulsen, 1973)
   males: 4.60–6.60 mm (Poulsen, 1973)

**Female (A–1).** Carapace rounded-oval in lateral view and large (4.60 mm long), with no indication of a cauda; height:length ratio about 0.75; rostrum short, overhanging and slightly flattened on top; incisure deep and U-shaped; a fringe of connected hyaline spines extends along the rostrum almost to the base of the incisure, ending in 2 hyaline spines, commences again on the opposite side of the incisure and extends for a short distance beyond the antero-ventral bulge. Under the rostral margin, the infold carries 8–9 spine-like bristles forming a vertical row, plus a clump of about 16 other bristles, mostly subequal in length, housed in a shallow subcircular "pocket" with a distinct inward rim, off which and facing inwards is a single hooked spine with a blunted tip; this clump of bristles is located near the inner margin of the incisure;

elsewhere, the infold is relatively wide antero-ventrally and ventrally, but narrows posteriorly; there is no posterior inner list, and the list below the rostrum is bare. Selvage lamellar and narrow, well developed antero-ventrally, ventrally and postero-ventrally. Colour yellowish. Surface smooth, but observed to have numerous simple normal pores. Adductor muscle scar pattern present medially but not well defined. Lateral eyes large and compound, reddish, each with about 35 ommatidia. Furca lamellar, carrying 8 claws, which decrease in strength and size from front to rear, all separated from the lamella, every one with a row of fine sawteeth on each side of its inner face.

**Male.** Carapace similar to that of the female. A1 with a strong sensory bristle bearing many filaments; 'b' and 'c' bristles having suctorial organs on their 3 proximal filaments; 'f' and 'g' bristles much longer than in the female and about twice the length of the carapace. A2 stronger than that of the female. Other limbs, the furca, upper lip, rod-shaped organ and lateral eyes all rather similar to those of the female (Poulsen 1973). Penis paired, proximal part large, distal process on each side finger-shaped and bent backwards; vas deferens unpaired (Müller 1906a).

### Remarks

This large predatory cypridinid was found only at St. 11 in the Straits of Magellan, where there were 6 individuals in 3 samples. Our specimens represent an A–1 female and A–2 female (sample 11/3), three A–3 females (sample 11/5), and a single A–8 individual (sample 11/7). The latter is a first free-living stage juvenile with reddish compound lateral eyes, proportionately well-developed antennule, antenna and mandible, and an arcuate upper lip without any glandular field. The last-named sample also carried several large but empty globular egg cases of this species. The stomach of the largest specimen contained decapod larvae and both calanoid and cyclopoid copepods, mostly ingested whole. The single specimen of Martens (1979) was an A4 juvenile. Poulsen (1962) records the length of adults as 4.8–6.6 mm (males) and 4.6–6.4 mm (females). We have found many more specimens in the latest samples from Magellan Straits.

The genus *Macrocypridina* is defined readily by its undivided upper lip which is semicircular, evenly rounded and serrated by numerous glandular openings ventrally, with some small postero-lateral processes; of these, the postero-ventral process is divided into 3 short tubes. Poulsen (1962) records several minor differences between adults of this species and the type (and only other described) species, *M. castanea* – well described by Skogsberg (1920) – noting that the main character of difference is the smaller number of distal cleaning bristles on the 7th limb in *M. poulseni* (9–13 dorso-distal and 12–14 ventro-distal bristles). However, in the new material we have a male in which the 7th limb has 18 dorso-distal and 17 ventro-distal bristles.

## Distribution

Poulsen (1962) gives the latitudinal range of this taxon as extending from 37°05′N to 46°58′S, in the Indian and Pacific Oceans, from near the east coast of South Africa to the Bay of Panama. The lone juvenile recorded by Martens (1979) came from off the coast of Chile at 37°10′S. Angel (1981) states that *M. castanea* is an upper mesopelagic species found in all oceans between 56°N and 50°S; this range and distribution probably includes some specimens of *M. poulseni*. Even so, our record is the most southerly yet recorded for the species, although Hartmann (1985) notes that it occurs in the Antarctic, but without giving locality details. According to Poulsen (1962), who had access to the most abundant material, the depth range of this species is from 50–4000 m. Martens' specimen was collected near the surface; he considers that the species characterizes Antiboreal Bottom and Subtropical Water (Martens 1979). Our six juveniles were collected at mesopelagic-shallow depths in the western part of the Straits (53°31′S, 72°38′W).

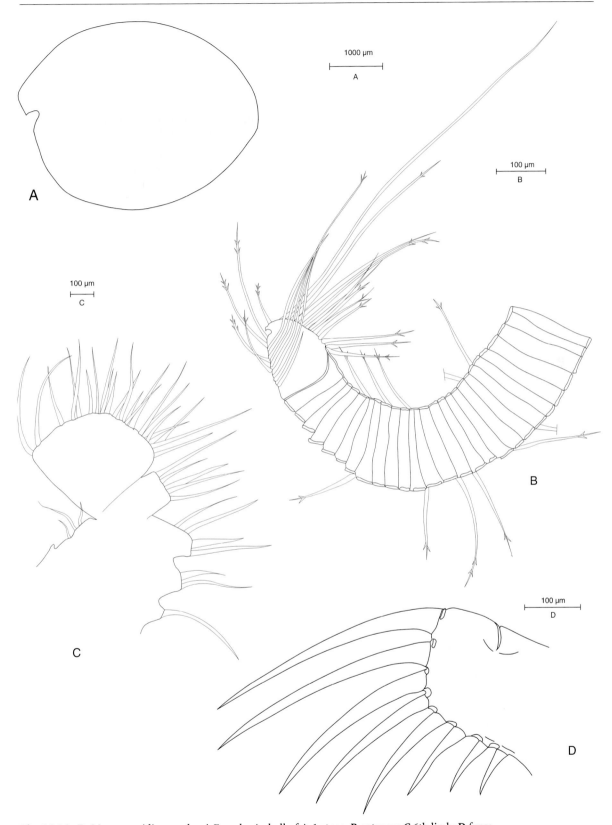

**Fig. 6.9.1A–D.** *Macrocypridina poulseni.* Female: **A** shell of A-1 stage; **B** antenna; **C** 6th limb; **D** furca

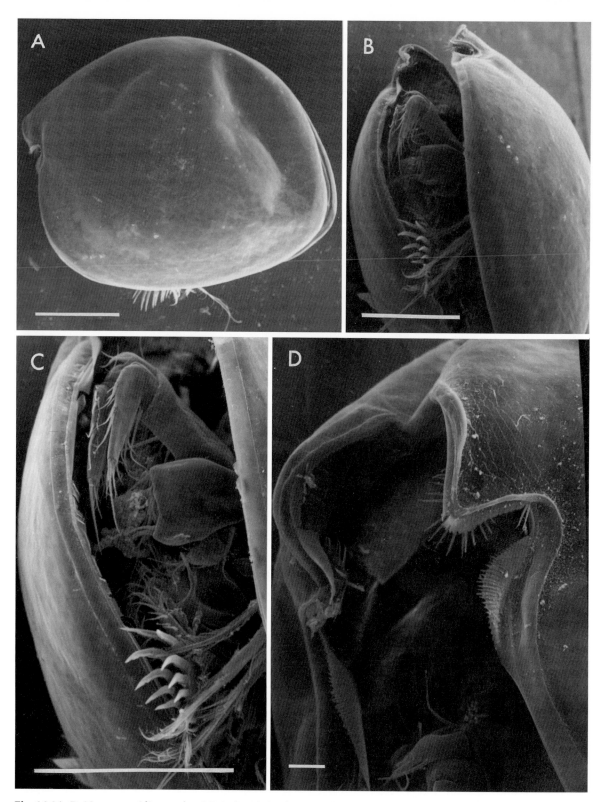

**Fig. 6.9.2A–D.** *Macrocyopridina poulseni.* Female: **A** left valve; **B,C** whole animal ventral view; **D** detail of the rostrum. *Bars* A,B,C 1 mm; D 100 μm

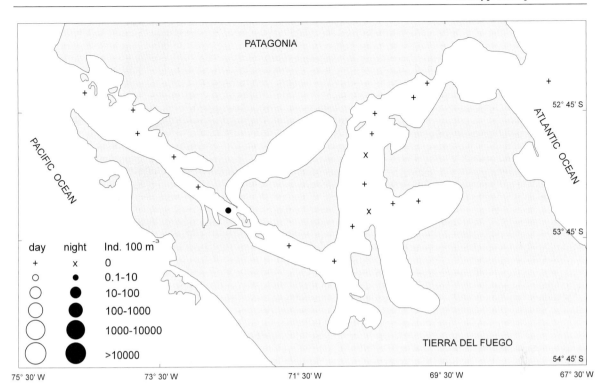

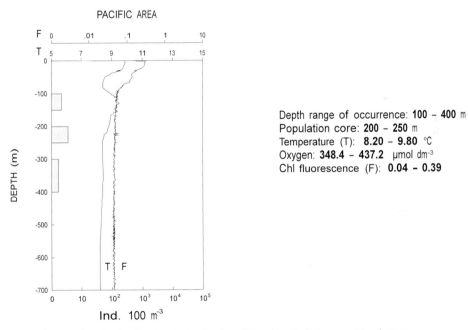

Depth range of occurrence: **100 – 400** m
Population core: **200 – 250** m
Temperature (T): **8.20 – 9.80** °C
Oxygen: **348.4 – 437.2** μmol dm⁻³
Chl fluorescence (F): **0.04 – 0.39**

**Fig. 6.9.3.** *Macrocyopridina poulseni.* Distribution in the Straits of Magellan in February–March 1991

## 6.10 *Philomedes eugeniae* Skogsberg, 1920

*Philomedes eugeniae* Skogsberg, 1920: pp. 410–413, fig. 74 – Poulsen, 1962: p. 345 (key) – Kornicker, 1975: pp. 238–240, figs. 135–138.

Carapace length –
  females: 1.37–1.75 mm (Kornicker, 1975)
  males: unknown

**Female.** Carapace medium-sized, with a weak subangular cauda; height about 2/3 the length; rostrum prominent, with a flattened top; incisure deep; dorsal margin strongly curved, obliquely truncate posterodorsally; ventral margin evenly rounded from weak cauda to the incisure. Medial bristles on the rostral row numerous; on the posterior list there are also some bristles, partly arranged in groups; between the posterior list and the valve margin are a few more bristles; rostral selvage bearing short marginal hairs. Carapace sculpture consisting of numerous subcircular and shallow pits. Colour whitish. 7th limb with 3 dorsal and 2 ventral distal bell bristles, plus a total of 9–13 proximal bell bristles. Furca lamellar with 10 claws decreasing in length and strength from front to rear.

**Male.** Unknown for this species.

## Remarks

We found five juveniles of this species. They are referred to *P. eugeniae* because of the posterodorsal profile of the shell and the weak cauda, the presence of numerous shallow pits on the valves, and the fact that, allowing for incremental growth of around 25 % at each ecdysis, their adult size would be within the size range given by Kornicker (1975, see above). Our specimens were collected at St. 21 (52°52′S, 70°30′W) and St. 22 (52°40′S, 69°55′W). The lateral profile of *P. eugeniae* is more truncated postero-dorsally than that of *P. assimilis* Brady, 1907 and the shell has scattered shallow, subcircular pits rather than the numerous punctae which characterize that species (Kornicker 1975, figs. 129c, 138b). Another, and larger, individual has turned up in the new samples from Magellan Straits.

## Distribution

Our specimens were taken in samples from the eastern end of Magellan Straits. Kornicker (1975) gives the type locality for this species as the Straits of Magellan at a depth of 7 m – material collected by the Swedish SS *Eugeniae* Expedition. This was the advice he received from the curator of the type specimens at the Naturhistoriska Riksmuseet, Stockholm. The paratypes were also collected in the same area, off Cape Valentyn, at 270 m. Other material, examined by Kornicker personally, was collected in Magellan Straits by the RV *Hero* cruises 69–5 and 70–2. Indeed, this species is known only from the Magellanic subregion of the sub-Antarctic, whereas *P. assimilis*, which has a similar shape but different surface sculpture, is confined to the Antarctic, at depths from 9 to 876 m (Kornicker 1975).

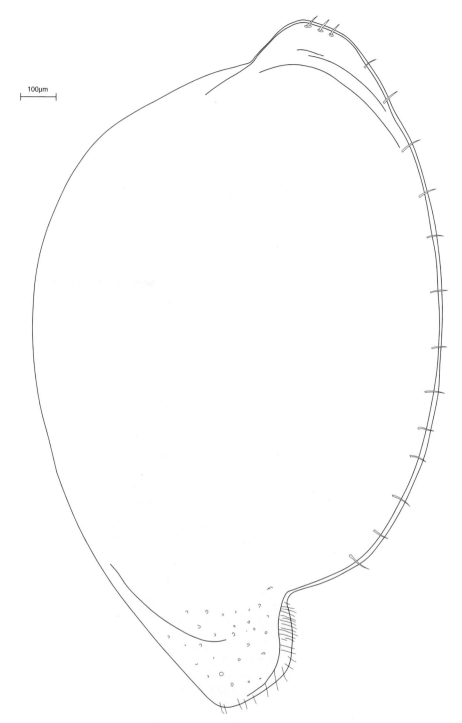

100µm

**Fig. 6.10.1.** *Philomedes eugeniae.* Female shell (juvenile?)

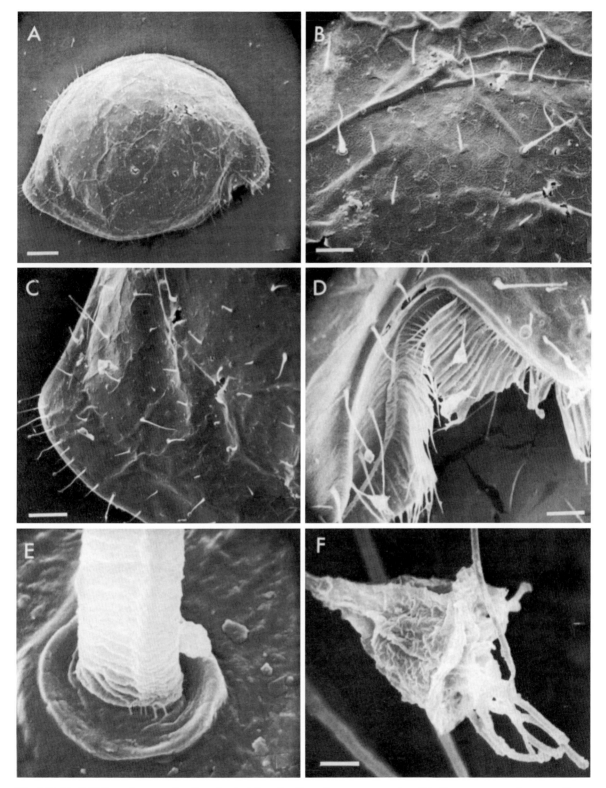

**Fig. 6.10.2A–F.** *Philomedes eugeniae.* Female: **A** right valve; **B** surface near posterior of **A;** **C** postero-ventral process; **D** anterior incisure; **E** detail of bristle pore; **F** ectoparasite shown in **D.** *Bars* **A** 200 μm; **B,D** 34 μm; **C** 60 μm; **E** 1 μm; **F** 3 μm. (Excerpt from Kornicker, 1975: p. 239, fig. 138)

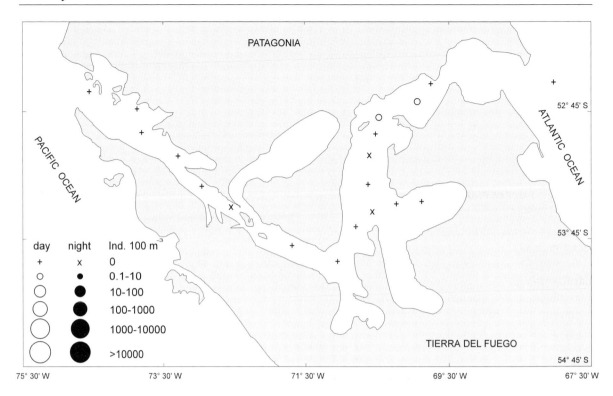

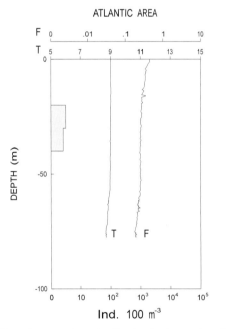

Depth range of occurrence: **20 – 40** m
Population core: **20 – 30** m
Temperature (T): **8.68 – 9.02** °C
Oxygen: **390.4 – 410.9** µmol dm⁻³
Chl fluorescence (F): **0.16 – 0.44**

**Fig. 6.10.3.** *Philomedes eugeniae*. Distribution in the Straits of Magellan in February–March 1991

## 6.11 *Philomedes cubitum* Kornicker, 1975

*Philomedes cubitum* Kornicker, 1975: pp. 289–293, figs. 132, 178, 179.

Carapace length –
  females: 1.39–1.40 mm (Kornicker, 1975)
  males: 1.59 mm

**Female.** Carapace mediumsized, subovate; height about 2/3 the length; rostrum curved and overhanging, with a flattened tip; incisure narrow and deep; no trace of a cauda; dorsal margin more or less evenly rounded but steeply truncate in the rear; ventral margin evenly rounded from the broadly subacuminate posterior to the incisure. Infold broad; rostral infold carrying about 21 bristles; anterior infold finely striate with about 9 short bristles; postero-ventral and posterior infold bearing about 25 bristles; selvage forming a right-angled elbow on the postero-ventral infold of both valves (species character). A1 sensory bristle without long filaments. A2 endopod 2-jointed. 7th limb with 3 dorsal and 2 ventral distal bell bristles, plus a total of 17 proximal bell bristles. Furca lamellar, with 5 stout followed by 5 weaker claws. Lateral eyes minute, each with 2 ommatidia; medial eye large. Upper lip hirsute, bearing a single elongate process. Rod-shaped organ long and slender, 2-jointed, with a rounded tip (Kornicker 1975).

**Male.** Carapace similar to that of the female except that the incisure is more open. Al sensory with long filaments, 'c' and 'f' bristles extremely long. A2 endopod 3-jointed. Maxilla reduced compared with that of the female.

### Remarks

Kornicker (1975) described only the female and an N–1 male of this species, because there were no mature males in the RV *Hero* cruise 69–5 type material, nor did the species occur in any other collections which he examined. Our only specimen, recognized by the elbow-like bend in the posterior selvage as belonging to *P. cubitum*, was a mature male close to the size limit set by Kornicker (1975) for the species. This unique specimen has been donated to the United States Museum of Natural History: Smithsonian Institution, Washington, D.C., USA (USNM Catalogue No. 194145). Unfortunately, because of the thin shell, attempts to carry out SEM photography of the specimen were unsuccessful (Kornicker, pers. comm.). It is of interest to add that the Antarctic and Subantarctic are areas of greatest diversity for the genus *Philomedes* (Kornicker 1975, figs. 132, 135).

### Distribution

The material described by Kornicker (1975) was collected by RV *Hero* at two stations in the Straits of Magellan (53°32′S, 72°29′W, and 53°41′ 40′′S, 72°45′′W) and from depths of 21.3–28.0 m. Our mature male was collected at St. 21 (52° 52′S, 70° 30′W). Since there are no other records, the species appears to be confined to shallow depths in the Straits and the Magellanic subregion of the sub-Antarctic, as defined by Kornicker (1975).

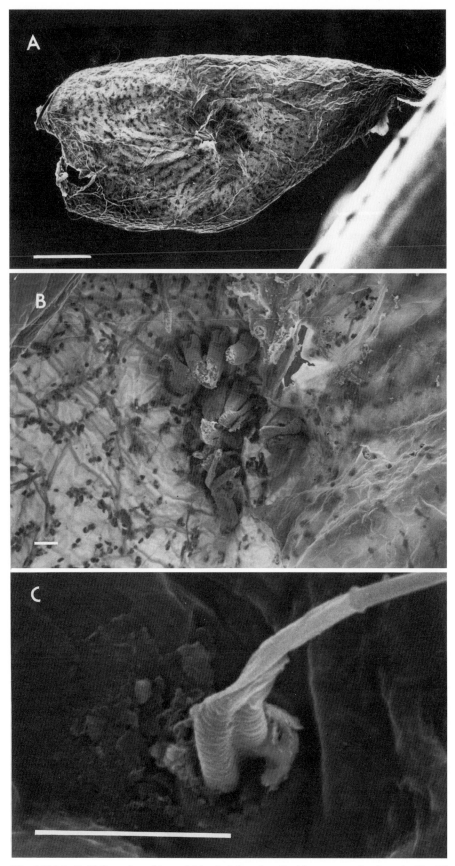

**Fig. 6.11.1A–C.** *Philomedes cubitum.* Male: **A** right valve; **B** adductor muscles from inside; **C** hair base and pore. *Bars* **A** 200 μm; **B** 20 μm; **C** 10 μm

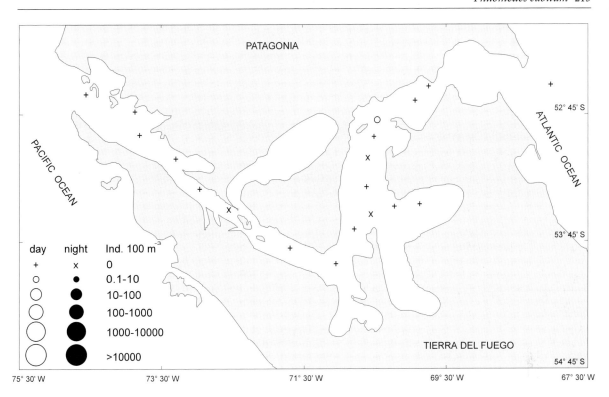

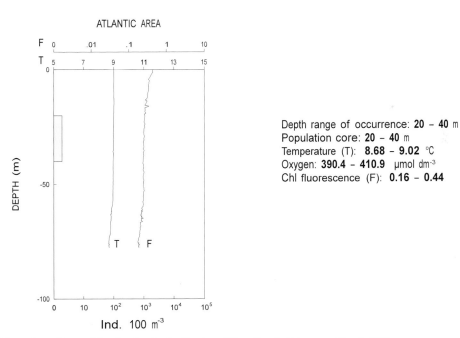

Depth range of occurrence: **20 – 40** m
Population core: **20 – 40** m
Temperature (T): **8.68 – 9.02** °C
Oxygen: **390.4 – 410.9** µmol dm⁻³
Chl fluorescence (F): **0.16 – 0.44**

**Fig. 6.11.2.** *Philomedes cubitum.* Distribution in the Straits of Magellan in February–March 1991

## 6.12 *Paradoxostoma* sp. aff. *hypselum* Müller, 1908

*Paradoxostoma hypselum* Müller, 1908: pp. 118–119, with 3 textfigs – Benson, 1964: p. 12, figs. 4, 5, pl. 1 fig. 2 – Neale, 1967: p. 9, pl. 1 figs g, k – McKenzie, 1972a: p. 161 – Kornicker, 1975: p. 235 – Hartmann, 1986: p. 172 – Hartmann, 1987: p. 131 – Hartmann, 1989: p. 250, figs. 61, 62 – Hartmann, 1990: p. 210.

Carapace length –
     females: 0.70–0.73 mm (Müller, 1908)
     males: 0.70–0.73 mm (Müller, 1908)

**Female.** Carapace small-medium (for a podocopid), shape subrhomboid; height:length ratio 0.55–0.60, greatest height posteromedial; dorsal margin strongly arched; ventral margin curved upwards posteriorly to make a weak cauda above the midline with the postero-dorsal edge, but slightly inflexed anteromedially; anterior more narrowly rounded, trending antero-ventrally; surface smooth; colour brownish-yellowish. Inner lamella broad, line of concrescence submarginal, vestibule large, extending from front to rear; marginal pore canals few and short; normal canals scattered, simple; central adductors in a subvertical row of 4, set medially; hinge adont. All limbs elongate and slender. A1 6-segmented; A2 5-segmented (including protopod), flagellum reaching beyond the terminal claws, with a large spinneret gland proximally; mandible coxale styliform, palp reduced, 2-segmented; oral cone with a suctorial apparatus (generic character); maxillule epipod with 14 strahlen, plus 2 downwards-pointing setae, palp missing, 3 lobes slender; P1 pediform, protopod armed with a powerful dorsodistal unguiform spine (generic charac-

ter); P2 and P3 also pediform but carrying bristles only on the protopods; posterior of the body lobate, with a spinulose tip; eye large, located anterodorsally.

**Male.** Similar in many respects to the female, except that the posterior of the body is not as produced. Hemipenis with a large subovate posterior capsule, anterior process broadly triangular, wedge-shaped.

### Remarks

This species was identified in many samples, including some in which planktonic ostracods predominated; e.g. we found 64 specimens at St. 22 (52°40′S, 69°55′W). In a comment on symbiosis among Ostracoda, McKenzie (1972a) noted that he had identified as *Paradoxostoma* cf. *hypselum* a specimen forwarded by Kornicker which had been removed from inside the shell of the myodocopid *Philomedes assimilis* and near, "…the area of muscle of the protopodite of the second antenna…" (L.S. Kornicker, pers. comm. 16 December 1971). McKenzie (1972a) recorded further that this individual when dissected had what appeared to be a piece of tissue attached near its suctorial apparatus, with its styliform mandibles both oriented correctly for suctorial feeding, i.e. aligned near the mouth. This suggests that some paradoxostomatids may parasitize epibenthic myodocopids. In the Magellan Straits samples collected by the Italian Antarctic Expedition, however, none of the few philomedids proved to have *P. hypselum* within their shells, so McKenzie's interesting earlier interpretation could not be tested further. Our abundant material of this taxon is referred only tentatively to *P. hypselum* because, although the shell shape is remarkably similar and the hemipenis seems nearly identical, the size of our specimens is considerably smaller (carapace length around 0.50 mm). Another similar species is *Paradoxostoma kensleyi* McKenzie 1972, described from near Cape Town, South Africa. Of about the same size (carapace length 0.52 mm) and with a similar hemipenis, it differs somewhat in shape (McKenzie 1972b).

## Distribution

This is the most widespread paradoxostomatid in Antarctic nearshore waters being reported from the Gauss Station (about 90°E), McMurdo Sound in the Ross Sea, Elephant Island in the South Shetlands, and Halley Bay in Coats Land (about 27°W). Since it has also been identified as a fossil in the Late Quaternary deposits of Watts Lake, Vestfold Hills, Australian Antarctic Territory (Gou and Li 1985), it can be assumed that its provenance is circum-Antarctic. Hartmann (1986, 1987, 1989, 1990) recorded it regularly from around Elephant Island, which is virtually opposite the Magellan Straits across the Drake Passage.

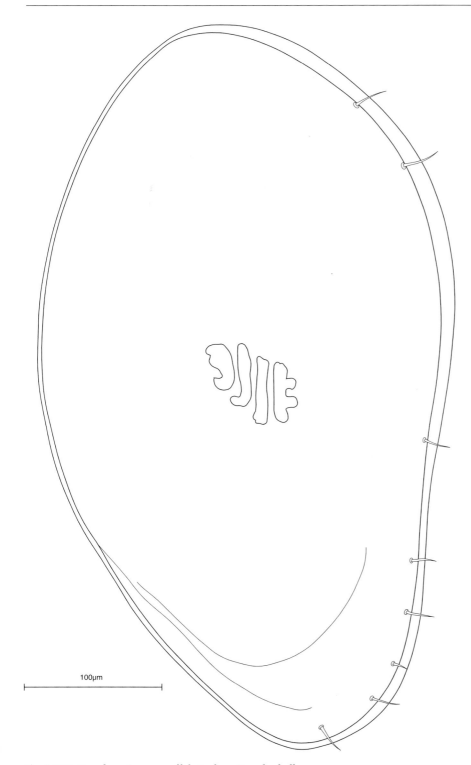

100µm

**Fig. 6.12.1.** *Paradoxostoma* sp. aff. *hypselum*. Female shell

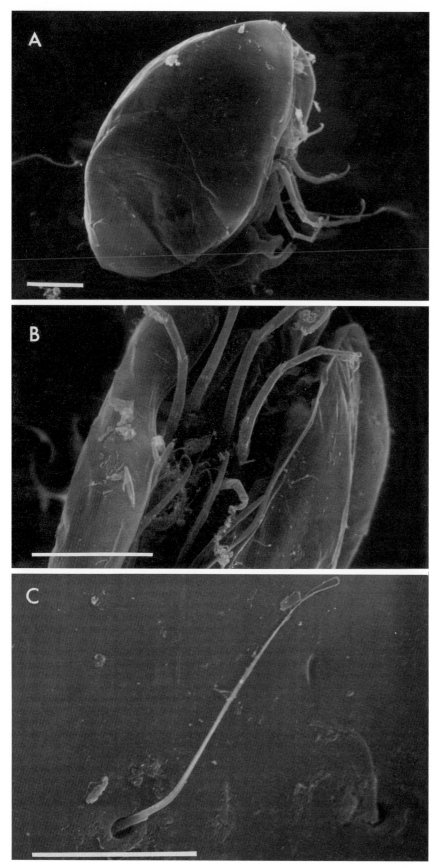

**Fig. 6.12.2A–C.** *Paradoxostoma* sp. aff. *hypselum*. Female: **A** right valve; **B** whole animal ventral view; **C** detail of bristle. *Bars* **A,B** = 100 µm; **C** = 10 µm

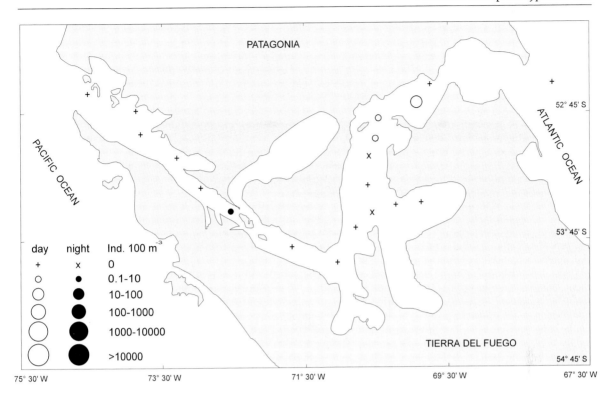

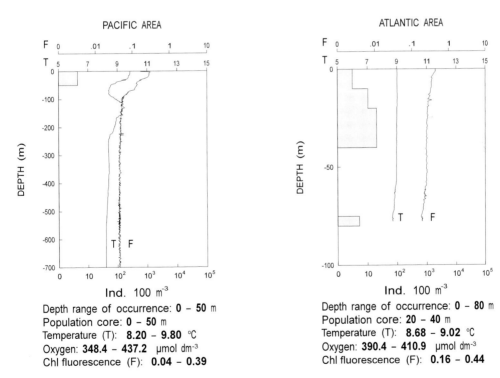

PACIFIC AREA

Depth range of occurrence: **0 – 50** m
Population core: **0 – 50** m
Temperature (T): **8.20 – 9.80** °C
Oxygen: **348.4 – 437.2** μmol dm⁻³
Chl fluorescence (F): **0.04 – 0.39**

ATLANTIC AREA

Depth range of occurrence: **0 – 80** m
Population core: **20 – 40** m
Temperature (T): **8.68 – 9.02** °C
Oxygen: **390.4 – 410.9** μmol dm⁻³
Chl fluorescence (F): **0.16 – 0.44**

**Fig. 6.12.3.** *Paradoxostoma* sp. aff. *hypselum*. Distribution in the Straits of Magellan in February–March 1991

## 6.13 *Paradoxostoma magellanicum* Müller, 1908

*Paradoxostoma magellanica* Müller, 1906b: p. 7, figs. 12–15.
*Paradoxostoma magellanicum* Müller, 1912: p. 287.

Carapace length –
   females: 0.56 mm
   males: 0.65 mm (Müller, 1906b)

**Female.** Carapace mediumsized, subelliptical; height a little more than 1/2 the length, greatest height about medial; dorsum gently arched, anterior more narrowly rounded than the broadly rounded posterior; ventral margin almost straight, but weakly inflexed anteromedially. Surface smooth; colour brownish. Inner lamella broad, line of concrescence submarginal anteriorly, displaced further inwards ventrally and posteriorly; vestibule large, broad anteriorly, elongate posteriorly; 15–20 marginal pore canals. Normal pore canals scattered simple. Central adductors comprising a subvertical row of 4 or 5 scars, several of them divided; hinge adont. Eye anterodorsal. Soft anatomy generally similar to that of *P.* sp. aff. *hypselum*.

**Male.** Similar in shape and soft anatomy to the female, except that the 3rd segment of the A1 is strongly hirsute; the maxillular palp is present but represented by only a simple bristle, and the first lobe of the maxillule is only 1/2 the length of lobes 2 and 3. The anterior process of the hemipenis is lanceolate (Müller 1906b).

### Remarks

We identified only 5 individuals as belonging to this species and none was male, so we were unable to check if the hemipenis matched the type description. This is a disappointing result, as the Straits of Magellan are the type locality for this species and it has been described from nowhere else. Interestingly, the single male on which the type description is based was found associated with the shell of *Eurypodius latreillei* (Müller 1906b).

### Distribution

Known only from the Straits of Magellan.

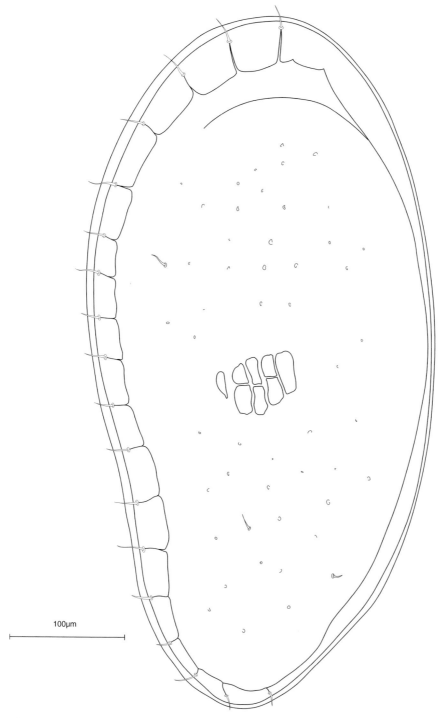

100µm

**Fig. 6.13.1.** *Paradoxostoma magellanicum.* Female shell

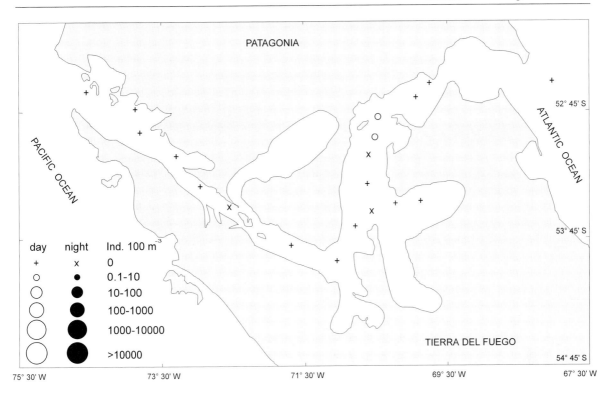

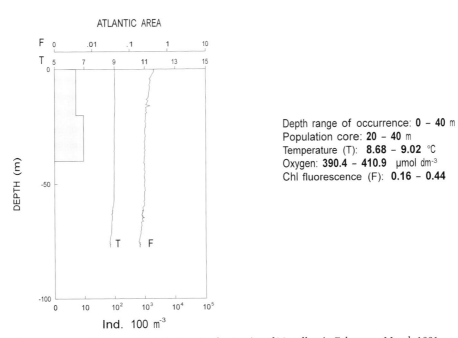

Depth range of occurrence: **0 – 40** m
Population core: **20 – 40** m
Temperature (T): **8.68 – 9.02** °C
Oxygen: **390.4 – 410.9** μmol dm⁻³
Chl fluorescence (F): **0.16 – 0.44**

**Fig. 6.13.2.** *Paradoxostoma magellanicum.* Distribution in the Straits of Magellan in February–March 1991

Angel MV (1972) Planktonic oceanic ostracods – historical, present and future. Proc R Soc Edinb (B) 73: 213–228

Angel MV (1981) Ostracoda. In: Boltovskoy D(ed) Atlas del zooplancton del Atlantico Sudoccidental y metodos de trabajo con zooplancton marino. Publicación (INIDEP), Ministerio de Comercio e Interes Marítimos, Mar del Plata, Argentina, pp 1–936

Aurivillius CWS (1898) Vergleichende thiergeographische Untersuchungen über die Plankton-Fauna des Skageraks in den Jahren 1893–1897. Vet Akad Handl Stockholm 30, 3

Benassi G, Ferrari I, Gentile G, Menozzi P, McKenzie KG (1992) Planktonic Ostracoda in the Southern Ocean and in the Ross Sea: 1989–90 Campaign. Nat Sci Comm Antarct Ocean Camp 1989–90 Data Rep 2: 247–300

Benassi G, Ferrari I, Gentile G, Menozzi P, McKenzie KG (1994a) Ostracoda in zooplankton samples from the Straits of Magellan: 1991 cruise. Nat Sci Comm Antarct Magellan Cruise February-March 1991 Data Rep 3: 191–213

Benassi G, Ferrari I, Menozzi P, McKenzie KG (1994b) Planktic ostracodes from the Antarctic and Subantarctic collected by the 1989–1990 Italian Antarctic Expedition. Rec Austr Mus 46: 25–37

Benson RH (1964) Recent cytheracean ostracodes from McMurdo Sound and the Ross Sea, Antarctica. Univ Kans Paleontol Contrib 6: 1–36

Brady GS (1880) Report on the Ostracoda dredged by H.M.S. Challenger during the years 1873–1876. Report of the Voyage of H.M.S. Challenger. Zool 1 (3): 1–184

Brady GS (1897) A supplementary report on the crustaceans of the group Myodocopa obtained during the Challenger expeditions with notes on other new or imperfectly known species. Trans Zool Soc Lond 14 (3)7: 85–100

Brady GS (1907) Crustacea V – Ostracoda. National Antarctic Expedition 1901–1904. Nat Hist Zool Bot Lond 3: 1–9

Chavtur VG (1993) The state of knowledge of recent myodocopid ostracodes of the world's oceans. In: McKenzie KG, Jones PJ (eds) Ostracoda in the earth and life sciences. Balkema, Rotterdam, pp 623–632

Claus C (1874a) Die Gattungen und Arten der Halocypriden. Verh Zool Bot Ges Wien 24: 175–178

Claus C (1874b) Die Familie der Halocypriden. Schr Zool Inhalts (Wien) 1: 1–16

Claus C (1890) Die Gattungen und Arten der mediterranen und atlantischen Halocypriden nebst Bemerkungen über die Organisation derselben. Arb Zool Inst Wien 9 (1): 1–34

Claus C (1891) Die Halocypriden des Atlantischen Oceans und Mittelmeeres. Arb Zool Inst Wien 4: 1–82

Claus C (1894) Zoologische Ergebnisse III. Die Halocypriden und ihre Entwicklungsstadien, gesammelt 1890, 1891, 1892, 1893. Denkschr Kais Akad Wiss Math-Nat Classe Wien 61: 1–10

Deevey GB (1978) A taxonomic and distributional study of the planktonic ostracods collected on three cruises of the Eltanin in the South Pacific and Antarctic region of the South Pacific. In: Pawson DL, Kornicker LS (eds) Biology of the Antarctic Seas VIII. Antarct Res Ser 28: 43–70

Deevey GB (1982) A faunistic study of the planktonic ostracods (Myodocopa, Halocyprididae) collected on eleven cruises of the Eltanin between New Zealand, Australia, the Ross Sea, and the South Indian Ocean. In: Kornicker LS (ed) Biology of the Antarctic Seas X. Antarct Res Ser 32: 131–167

Deevey GB (1983) Planktonic ostracods (Myodocopa, Halocyprididae) from six Eltanin cruises in South Pacific and Antarctic waters. J Crustacean Biol 3 (3): 409–416

Gooday A (1981) The Conchoecia skogsbergi species complex (Ostracoda, Halocyprididae) in the Atlantic Ocean. Bull Br Mus Nat Hist (Zool) 40 (4): 137–209

Gou Y, Li Y (1985) The Holocene Ostracoda from Section DWI of Lake Watts in the Vestfold Hills, Antarctica. Collected Papers on Antarctic Scientific Research, Studies of Late Quaternary Geology and Geomorphology in the Vestfold Hills, east Antarctica. Science Press, Beijing (in Chinese)

Hartmann G (1985) Ostracoden aus der Tiefsee des Indischen Ozeans und der Iberischen See sowie von ostatlantischen sublitoralen Plateaus und Kuppen. Mit einer Tabelle der bislang bekannten rezenten Tiefseeostracoden. Senckenb Marit 17 (1/3): 89–146

Hartmann G (1986) Antarktische benthische Ostracoden I. Mit einer Tabelle der bislang aus der Antarktis bekannten Ostracoden. Mitt Hamb Zool Mus Inst 83: 147–221

Hartmann G (1987) Antarktische benthische Ostracoden II. Auswertung der Fahrten der Polarstern Ant. III /2 und der Reisen der Walther Herwig 68/1 und 2. 2 Teil: Elephant Island und Bransfield Strasse. Mitt Hamb Zool Mus Inst 84: 115–156

Hartmann G (1989) Antarktische benthische Ostracoden V. Auswertung der Südwinterreise von FS Polarstern (Ps 9/V-1) im Bereich Elephant Insel und der Antarktischen Halbinsel. Mitt Hamb Zool Mus Inst 86: 231–288

Hartmann G (1990) Antarktische benthische Ostracoden VI. Auswertung der Fahrten der Reise der Polarstern ANT VI-2 (1. Teil, Meiofauna und Zehnerserien) sowie Versuch einer vorläufigen Auswertung aller bislang vorliegenden Daten. Mitt Hamb Zool Mus Inst 87: 191–245

Iles EJ (1953) A preliminary report on the Ostracoda of the

Benguela Current. Discovery Rep 26: 259–280

Koch R (1992) Ostracoden im Epipelagial vor der Antarktischen Halbinsel – ein Beitrag zur Systematik sowie zur Verbreitung und Populationsstruktur unter Berücksichtigung der Saisonalität. Ber Polarforsch 106: 1–209

Kornicker LS (1975) Antarctic Ostracoda (Myodocopina). Part 1. Smithson Contrib Zool 163: 1–374

Lubbock J (1860) On some oceanic Entomostraca collected by Captain Toynbee. Trans Linn Soc Lond 23: 173–191

Martens JM (1979) Die pelagischen Ostracoden der Expedition *Marchile* I (Südost-Pazifik) II: Systematik und Vorkommen (Crustacea: Ostracoda: Myodocopida). Mitt Hamb Zool Mus Inst 76: 303–366

McKenzie KG (1972a) New data on the ostracode genera *Laocoonella* de Vos & Stock, *Redekea* and *Aspidoconcha* de Vos, with a key to the family Xestoleberididae and a rèsumé of symbiosis in Ostracoda. Beaufortia 19: 151–162

McKenzie KG (1972b) A new species of *Paradoxostoma* (Crustacea, Ostracoda) from South Africa. Ann S Afr Mus 59: 133–137

McKenzie KG, Benassi G, Naldi M, Ferrari I, Menozzi P (1990) Report on planktic Ostracoda from the Ross Sea, Antarctica. Nat Sci Comm Antarct Ocean Camp 1987–88 Data Rep 2: 171–229

Müller GW (1906a) Ostracoda. Wiss Ergeb Dtsch Tiefsee-Exped *Valdivia* 1898–1899, 8: 29–154

Müller GW (1906b) Ostracoden. Résultats du voyage du S.Y.B. *Belgica* en 1897–1898–1899. Rapp Sci Zool: 1–10

Müller GW (1908) Die Ostracoden der Deutschen Südpolar-Expedition 1901–1903. Wiss Ergeb Dtschn Südpolar Expedition, Bd 10, Zoology, II Bd (Berl), pp 58–181

Müller GW (1912) Ostracoda. In: Schulze FE (ed) Das Tierreich. Auftrage Konigl Preuss, Akad Wiss (Berl) 31, pp 1–434

Neale JW (1967) An ostracod fauna from Halley Bay, Coats Land, British Antarctic Territory. Br Antarct Surv Sci Rep 58: 1–50

Poulsen EM (1962) Ostracoda-Myodocopa Part I Cypridiniformes-Cyprinidae. Dana Rep 57: 1–414

Poulsen EM (1973) Ostracoda-Myodocopa Part III B Halocypriformes-Halocypridae Conchoecinae. Dana Rep 84: 1–224

Ramirez FC, Moguilevsky A (1971) Ostracodos planctonicos hallados en aguas oceanicas frente a la provincia de Buenos Aires (Resultados de la XLI Comisao Oceanografica Costa Sul). Physis 30 (81): 637–666

Sars GO (1865) Oversigt of Norges marine Ostracoder. Forh Vid-Selsk Christiania: 1–130

Skogsberg T (1920) Studies on marine ostracods. Part I. Cypridinids, halocyprids and polycopids. Zool Bidr Uppsala Suppl 1: 1–784

## Stations and Sample Data for Ostracods During R/V *Cariboo* Cruise in the Straits of Magellan

*Mikroconchoecia cf. acuticosta*

| Station number | Depth (m) | Ind·m$^{-3}$ | Station number | Depth (m) | Ind·m$^{-3}$ | Station number | Depth (m) | Ind·m$^{-3}$ |
|---|---|---|---|---|---|---|---|---|
| 5 | 160–140 | 0.00 | 13 | 400–300 | 0.00 | 19 | 100–75 | 0.00 |
|  | 115–100 | 0.00 |  | 250–200 | 0.00 |  | 75–50 | 0.00 |
|  | 80–60 | 0.00 |  | 150–100 | 0.00 |  | 50–25 | 0.00 |
|  | 40–20 | 0.00 |  | 50–25 | 0.00 |  | 25–0 | 0.00 |
| 6 | 455–400 | 0.00 | 14 | 400–300 | 0.00 | 20 | 80–75 | 0.04 |
|  | 300–250 | 0.00 |  | 250–200 | 0.00 |  | 60–40 | 0.00 |
|  | 200–150 | 0.00 |  | 150–100 | 0.00 |  | 40–20 | 0.00 |
|  | 100–50 | 0.00 |  | 90–50 | 0.00 |  | 20–0 | 0.00 |
| 7 | 500–400 | 0.00 | 15 | 300–200 | 0.00 | 21 | 40–20 | 0.01 |
|  | 300–250 | 0.02 |  | 150–100 | 0.00 |  | 20–0 | 0.00 |
|  | 200–150 | 0.00 |  | 80–60 | 0.00 |  |  |  |
|  | 100–50 | 0.02 |  | 40–20 | 0.00 |  |  |  |
| 9 | 700–600 | 0.00 | 16 | 80–60 | 0.04 | 22 | 30–20 | 0.00 |
|  | 400–300 | 0.00 |  | 60–40 | 0.00 |  | 20–10 | 0.00 |
|  | 200–140 | 0.00 |  | 40–20 | 0.00 |  | 10–0 | 0.00 |
|  | 100–50 | 0.00 |  | 20–0 | 0.00 |  |  |  |
| 10 | 500–400 | 0.00 | 17 | 120–100 | 0.00 | 23 | 30–0 | 0.00 |
|  | 300–250 | 0.00 |  | 80–60 | 0.00 |  |  |  |
|  | 200–150 | 0.00 |  | 40–30 | 0.00 |  |  |  |
|  | 100–50 | 0.00 |  | 20–10 | 0.00 |  |  |  |
| 11 | 400–300 | 0.00 | 18 | 160–120 | 0.01 | 26 | 30–20 | 0.00 |
|  | 250–200 | 0.00 |  | 100–80 | 0.05 |  | 20–0 | 0.00 |
|  | 150–100 | 0.00 |  | 60–40 | 0.00 |  |  |  |
|  | 50–25 | 0.00 |  | 20–10 | 0.00 |  |  |  |
| 12 | 130–100 | 0.00 |  |  |  |  |  |  |
|  | 100–50 | 0.00 |  |  |  |  |  |  |
|  | 50-25 | 0.00 |  |  |  |  |  |  |
|  | 25–0 | 0.00 |  |  |  |  |  |  |

*Discoconchoecia elegans*

| Station number | Depth (m) | Ind·m$^{-3}$ | Station number | Depth (m) | Ind·m$^{-3}$ | Station number | Depth (m) | Ind·m$^{-3}$ |
|---|---|---|---|---|---|---|---|---|
| 5 | 160–140 | 0.00 | 13 | 400–300 | 18.25 | 19 | 100–75 | 27.09 |
|   | 115–100 | 0.00 |    | 250–200 | 111.76 |    | 75–50 | 10.37 |
|   | 80–60 | 0.00 |    | 150–100 | 27.88 |    | 50–25 | 4.43 |
|   | 40–20 | 0.00 |    | 50–25 | 5.48 |    | 25–0 | 0.69 |
| 6 | 455–400 | 0.61 | 14 | 400–300 | 167.86 | 20 | 80–75 | 12.90 |
|   | 300–250 | 0.33 |    | 250–200 | 202.14 |    | 60–40 | 3.08 |
|   | 200–150 | 0.57 |    | 150–100 | 70.14 |    | 40–20 | 2.65 |
|   | 100–50 | 0.01 |    | 90–50 | 6.25 |    | 20–0 | 0.05 |
| 7 | 500–400 | 0.59 | 15 | 300–200 | 91.19 | 21 | 40–20 | 3.21 |
|   | 300–250 | 0.50 |    | 150–100 | 76.19 |    | 20–0 | 1.57 |
|   | 200–150 | 0.31 |    | 80–60 | 126.29 |    |    |    |
|   | 100–50 | 0.02 |    | 40–20 | 426.01 |    |    |    |
| 9 | 700–600 | 0.13 | 16 | 80–60 | 40.20 | 22 | 30–20 | 0.00 |
|   | 400–300 | 0.37 |    | 60–40 | 7.58 |    | 20–10 | 0.09 |
|   | 200–140 | 4.29 |    | 40–20 | 3.57 |    | 10–0 | 0.07 |
|   | 100–50 | 0.56 |    | 20–0 | 0.44 |    |    |    |
| 10 | 500–400 | 0.91 | 17 | 120–100 | 630.64 | 23 | 30–0 | 0.02 |
|   | 300–250 | 4.35 |    | 80–60 | 18.04 |    |    |    |
|   | 200–150 | 7.58 |    | 40–30 | 9.49 |    |    |    |
|   | 100–50 | 1.02 |    | 20–10 | 10.63 |    |    |    |
| 11 | 400–300 | 1.08 | 18 | 160–120 | 436.47 | 26 | 30–20 | 0.00 |
|   | 250–200 | 7.36 |    | 100–80 | 20.80 |    | 20–0 | 0.02 |
|   | 150–100 | 7.17 |    | 60–40 | 11.08 |    |    |    |
|   | 50–25 | 12.82 |    | 20–10 | 14.42 |    |    |    |
| 12 | 130–100 | 13.56 |    |    |    |    |    |    |
|   | 100–50 | 2.28 |    |    |    |    |    |    |
|   | 50–25 | 0.46 |    |    |    |    |    |    |
|   | 25–0 | 0.18 |    |    |    |    |    |    |

*Loricoecia loricata*

| Station number | Depth (m) | Ind·m$^{-3}$ | Station number | Depth (m) | Ind·m$^{-3}$ | Station number | Depth (m) | Ind·m$^{-3}$ |
|---|---|---|---|---|---|---|---|---|
| 5 | 160–140 | 0.00 | 13 | 400–300 | 0.00 | 19 | 100–75 | 0.00 |
| | 115–100 | 0.00 | | 250–200 | 0.00 | | 75–50 | 0.00 |
| | 80–60 | 0.00 | | 150–100 | 0.00 | | 50–25 | 0.00 |
| | 40–20 | 0.00 | | 50–25 | 0.00 | | 25–0 | 0.00 |
| 6 | 455–400 | 0.00 | 14 | 400–300 | 0.00 | 20 | 80–75 | 0.00 |
| | 300–250 | 0.00 | | 250–200 | 0.00 | | 60–40 | 0.00 |
| | 200–150 | 0.00 | | 150–100 | 0.00 | | 40–20 | 0.00 |
| | 100–50 | 0.01 | | 90–50 | 0.00 | | 20–0 | 0.00 |
| 7 | 500–400 | 0.40 | 15 | 300–200 | 0.00 | 21 | 40–20 | 0.00 |
| | 300–250 | 0.50 | | 150–100 | 0.00 | | 20–0 | 0.00 |
| | 200–150 | 0.11 | | 80–60 | 0.00 | | | |
| | 100–50 | 0.00 | | 40–20 | 0.00 | | | |
| 9 | 700–600 | 0.00 | 16 | 80–60 | 0.00 | 22 | 30–20 | 0.00 |
| | 400–300 | 0.11 | | 60–40 | 0.00 | | 20–10 | 0.00 |
| | 200–140 | 0.05 | | 40–20 | 0.00 | | 10–0 | 0.00 |
| | 100–50 | 0.00 | | 20–0 | 0.00 | | | |
| 10 | 500–400 | 0.02 | 17 | 120–100 | 0.00 | 23 | 30–0 | 0.00 |
| | 300–250 | 0.38 | | 80–60 | 0.00 | | | |
| | 200–150 | 0.00 | | 40–30 | 0.00 | | | |
| | 100–50 | 0.00 | | 20–10 | 0.00 | | | |
| 11 | 400–300 | 1.60 | 18 | 160–120 | 0.00 | 26 | 30–20 | 0.00 |
| | 250–200 | 0.02 | | 100–80 | 0.00 | | 20–0 | 0.00 |
| | 150–100 | 0.00 | | 60–40 | 0.00 | | | |
| | 50–25 | 0.00 | | 20–10 | 0.00 | | | |
| 12 | 130–100 | 0.00 | | | | | | |
| | 100–50 | 0.00 | | | | | | |
| | 50–25 | 0.00 | | | | | | |
| | 25–0 | 0.00 | | | | | | |

*Paramollicia rhynchena*

| Station number | Depth (m) | Ind·m⁻³ | Station number | Depth (m) | Ind·m⁻³ | Station number | Depth (m) | Ind·m⁻³ |
|---|---|---|---|---|---|---|---|---|
| 5 | 160–140 | 0.00 | 13 | 400–300 | 0.00 | 19 | 100–75 | 0.00 |
|   | 115–100 | 0.00 |    | 250–200 | 0.00 |    | 75–50 | 0.00 |
|   | 80–60 | 0.00 |    | 150–100 | 0.00 |    | 50–25 | 0.00 |
|   | 40–20 | 0.00 |    | 50–25 | 0.00 |    | 25–0 | 0.00 |
| 6 | 455–400 | 0.12 | 14 | 400–300 | 0.01 | 20 | 80–75 | 0.00 |
|   | 300–250 | 0.03 |    | 250–200 | 0.00 |    | 60–40 | 0.00 |
|   | 200–150 | 0.00 |    | 150–100 | 0.00 |    | 40–20 | 0.00 |
|   | 100–50 | 0.00 |    | 90–50 | 0.00 |    | 20–0 | 0.00 |
| 7 | 500–400 | 2.23 | 15 | 300–200 | 0.00 | 21 | 40–20 | 0.00 |
|   | 300–250 | 0.00 |    | 150–100 | 0.00 |    | 20–0 | 0.00 |
|   | 200–150 | 0.05 |    | 80–60 | 0.00 |    |   |   |
|   | 100–50 | 0.04 |    | 40–20 | 0.00 |    |   |   |
| 9 | 700–600 | 0.14 | 16 | 80–60 | 0.00 | 22 | 30–20 | 0.00 |
|   | 400–300 | 3.17 |    | 60–40 | 0.00 |    | 20–10 | 0.00 |
|   | 200–140 | 0.14 |    | 40–20 | 0.00 |    | 10–0 | 0.00 |
|   | 100–50 | 0.00 |    | 20–0 | 0.00 |    |   |   |
| 10 | 500–400 | 0.31 | 17 | 120–100 | 0.00 | 23 | 30–0 | 0.00 |
|   | 300–250 | 0.07 |    | 80–60 | 0.00 |    |   |   |
|   | 200–150 | 0.03 |    | 40–30 | 0.00 |    |   |   |
|   | 100–50 | 0.00 |    | 20–10 | 0.00 |    |   |   |
| 11 | 400–300 | 2.40 | 18 | 160–120 | 0.00 | 26 | 30–20 | 0.00 |
|   | 250–200 | 0.03 |    | 100–80 | 0.00 |    | 20–0 | 0.00 |
|   | 150–100 | 0.00 |    | 60–40 | 0.00 |    |   |   |
|   | 50–25 | 0.00 |    | 20–10 | 0.00 |    |   |   |
| 12 | 130–100 | 0.00 |    |   |   |    |   |   |
|   | 100–50 | 0.05 |    |   |   |    |   |   |
|   | 50–25 | 0.00 |    |   |   |    |   |   |
|   | 25–0 | 0.00 |    |   |   |    |   |   |

*Metaconchoecia australis*

| Station number | Depth (m) | Ind·m$^{-3}$ | Station number | Depth (m) | Ind·m$^{-3}$ | Station number | Depth (m) | Ind·m$^{-3}$ |
|---|---|---|---|---|---|---|---|---|
| 5 | 160–140 | 0.00 | 13 | 400–300 | 0.00 | 19 | 100–75 | 0.00 |
|  | 115–100 | 0.00 |  | 250–200 | 0.00 |  | 75–50 | 0.00 |
|  | 80–60 | 0.00 |  | 150–100 | 0.00 |  | 50–25 | 0.00 |
|  | 40–20 | 0.00 |  | 50–25 | 0.00 |  | 25–0 | 0.00 |
| 6 | 455–400 | 0.40 | 14 | 400–300 | 0.00 | 20 | 80–75 | 0.00 |
|  | 300–250 | 3.80 |  | 250–200 | 0.00 |  | 60–40 | 0.00 |
|  | 200–150 | 16.36 |  | 150–100 | 0.00 |  | 40–20 | 0.00 |
|  | 100–50 | 0.02 |  | 90–50 | 0.00 |  | 20–0 | 0.00 |
| 7 | 500–400 | 0.17 | 15 | 300–200 | 0.00 | 21 | 40–20 | 0.00 |
|  | 300–250 | 3.16 |  | 150–100 | 0.00 |  | 20–0 | 0.00 |
|  | 200–150 | 0.68 |  | 80–60 | 0.00 |  |  |  |
|  | 100–50 | 4.93 |  | 40–20 | 0.00 |  |  |  |
| 9 | 700–600 | 0.00 | 16 | 80–60 | 0.00 | 22 | 30–20 | 0.00 |
|  | 400–300 | 0.11 |  | 60–40 | 0.00 |  | 20–10 | 0.00 |
|  | 200–140 | 1.98 |  | 40–20 | 0.00 |  | 10–0 | 0.00 |
|  | 100–50 | 0.15 |  | 20–0 | 0.00 |  |  |  |
| 10 | 500–400 | 0.00 | 17 | 120–100 | 0.00 | 23 | 30–0 | 0.00 |
|  | 300–250 | 0.00 |  | 80–60 | 0.00 |  |  |  |
|  | 200–150 | 0.20 |  | 40–30 | 0.00 |  |  |  |
|  | 100–50 | 0.00 |  | 20–10 | 0.00 |  |  |  |
| 11 | 400–300 | 0.01 | 18 | 160–120 | 0.00 | 26 | 30–20 | 0.00 |
|  | 250–200 | 0.63 |  | 100–80 | 0.00 |  | 20–0 | 0.00 |
|  | 150–100 | 0.36 |  | 60–40 | 0.00 |  |  |  |
|  | 50–25 | 0.00 |  | 20–10 | 0.00 |  |  |  |
| 12 | 130–100 | 0.44 |  |  |  |  |  |  |
|  | 100–50 | 0.00 |  |  |  |  |  |  |
|  | 50–25 | 0.00 |  |  |  |  |  |  |
|  | 25–0 | 0.03 |  |  |  |  |  |  |

*Conchoecilla chuni*

| Station number | Depth (m) | Ind·m$^{-3}$ | Station number | Depth (m) | Ind·m$^{-3}$ | Station number | Depth (m) | Ind·m$^{-3}$ |
|---|---|---|---|---|---|---|---|---|
| 5 | 160–140 | 0.00 | 13 | 400–300 | 0.00 | 19 | 100–75 | 0.00 |
|   | 115–100 | 0.00 |    | 250–200 | 0.00 |    | 75–50 | 0.00 |
|   | 80–60 | 0.00 |    | 150–100 | 0.00 |    | 50–25 | 0.00 |
|   | 40–20 | 0.00 |    | 50–25 | 0.00 |    | 25–0 | 0.00 |
| 6 | 455–400 | 0.00 | 14 | 400–300 | 0.00 | 20 | 80–75 | 0.00 |
|   | 300–250 | 0.00 |    | 250–200 | 0.00 |    | 60–40 | 0.00 |
|   | 200–150 | 0.00 |    | 150–100 | 0.00 |    | 40–20 | 0.00 |
|   | 100–50 | 0.00 |    | 90–50 | 0.00 |    | 20–0 | 0.00 |
| 7 | 500–400 | 0.00 | 15 | 300–200 | 0.00 | 21 | 40–20 | 0.00 |
|   | 300–250 | 0.00 |    | 150–100 | 0.00 |    | 20–0 | 0.00 |
|   | 200–150 | 0.00 |    | 80–60 | 0.00 |    |   |   |
|   | 100–50 | 0.00 |    | 40–20 | 0.00 |    |   |   |
| 9 | 700–600 | 0.00 | 16 | 80–60 | 0.00 | 22 | 30–20 | 0.00 |
|   | 400–300 | 0.00 |    | 60–40 | 0.00 |    | 20–10 | 0.00 |
|   | 200–140 | 0.00 |    | 40–20 | 0.00 |    | 10–0 | 0.00 |
|   | 100–50 | 0.00 |    | 20–0 | 0.00 |    |   |   |
| 10 | 500–400 | 0.01 | 17 | 120–100 | 0.00 | 23 | 30–0 | 0.00 |
|   | 300–250 | 0.00 |    | 80–60 | 0.00 |    |   |   |
|   | 200–150 | 0.00 |    | 40–30 | 0.00 |    |   |   |
|   | 100–50 | 0.00 |    | 20–10 | 0.00 |    |   |   |
| 11 | 400–300 | 0.00 | 18 | 160–120 | 0.00 | 26 | 30–20 | 0.00 |
|   | 250–200 | 0.00 |    | 100–80 | 0.00 |    | 20–0 | 0.00 |
|   | 150–100 | 0.00 |    | 60–40 | 0.00 |    |   |   |
|   | 50–25 | 0.00 |    | 20–10 | 0.00 |    |   |   |
| 12 | 130–100 | 0.00 |    |   |   |    |   |   |
|   | 100–50 | 0.00 |    |   |   |    |   |   |
|   | 50–25 | 0.00 |    |   |   |    |   |   |
|   | 25–0 | 0.00 |    |   |   |    |   |   |

*Pseudoconchoecia serrulata*

| Station number | Depth (m) | Ind·m$^{-3}$ | Station number | Depth (m) | Ind·m$^{-3}$ | Station number | Depth (m) | Ind·m$^{-3}$ |
|---|---|---|---|---|---|---|---|---|
| 5 | 160–140 | 0.67 | 13 | 400–300 | 0.00 | 19 | 100–75 | 0.00 |
|   | 115–100 | 1.12 |    | 250–200 | 0.00 |    | 75–50 | 0.00 |
|   | 80–60 | 1.44 |    | 150–100 | 0.00 |    | 50–25 | 0.00 |
|   | 40–20 | 0.64 |    | 50–25 | 0.00 |    | 25–0 | 0.00 |
| 6 | 455–400 | 0.00 | 14 | 400–300 | 0.00 | 20 | 80–75 | 0.00 |
|   | 300–250 | 0.03 |    | 250–200 | 0.00 |    | 60–40 | 0.00 |
|   | 200–150 | 0.05 |    | 150–100 | 0.03 |    | 40–20 | 0.00 |
|   | 100–50 | 0.09 |    | 90–50 | 0.00 |    | 20–0 | 0.00 |
| 7 | 500–400 | 0.00 | 15 | 300–200 | 0.00 | 21 | 40–20 | 0.00 |
|   | 300–250 | 0.00 |    | 150–100 | 0.00 |    | 20–0 | 0.00 |
|   | 200–150 | 0.00 |    | 80–60 | 0.00 |    |  |  |
|   | 100–50 | 0.00 |    | 40–20 | 0.00 |    |  |  |
| 9 | 700–600 | 0.00 | 16 | 80–60 | 0.00 | 22 | 30–20 | 0.00 |
|   | 400–300 | 0.00 |    | 60–40 | 0.00 |    | 20–10 | 0.00 |
|   | 200–140 | 0.00 |    | 40–20 | 0.00 |    | 10–0 | 0.00 |
|   | 100–50 | 0.00 |    | 20–0 | 0.00 |    |  |  |
| 10 | 500–400 | 0.00 | 17 | 120–100 | 0.00 | 23 | 30–0 | 0.00 |
|   | 300–250 | 0.00 |    | 80–60 | 0.00 |    |  |  |
|   | 200–150 | 0.00 |    | 40–30 | 0.00 |    |  |  |
|   | 100–50 | 0.00 |    | 20–10 | 0.00 |    |  |  |
| 11 | 400–300 | 0.00 | 18 | 160–120 | 0.00 | 26 | 30–20 | 0.23 |
|   | 250–200 | 0.00 |    | 100–80 | 0.00 |    | 20–0 | 0.02 |
|   | 150–100 | 0.00 |    | 60–40 | 0.00 |    |  |  |
|   | 50–25 | 0.00 |    | 20–10 | 0.00 |    |  |  |
| 12 | 130–100 | 0.00 |    |  |  |    |  |  |
|   | 100–50 | 0.00 |    |  |  |    |  |  |
|   | 50–25 | 0.00 |    |  |  |    |  |  |
|   | 25–0 | 0.00 |    |  |  |    |  |  |

*Obtusoecia antarctica*

| Station number | Depth (m) | Ind·m$^{-3}$ | Station number | Depth (m) | Ind·m$^{-3}$ | Station number | Depth (m) | Ind·m$^{-3}$ |
|---|---|---|---|---|---|---|---|---|
| 5 | 160–140 | 0.24 | 13 | 400–300 | 0.00 | 19 | 100–75 | 0.05 |
|   | 115–100 | 0.46 |   | 250–200 | 0.00 |   | 75–50 | 0.00 |
|   | 80–60 | 0.20 |   | 150–100 | 0.05 |   | 50–25 | 0.00 |
|   | 40–20 | 0.07 |   | 50–25 | 0.00 |   | 25–0 | 0.00 |
| 6 | 455–400 | 0.30 | 14 | 400–300 | 0.03 | 20 | 80–75 | 0.09 |
|   | 300–250 | 2.20 |   | 250–200 | 0.09 |   | 60–40 | 0.00 |
|   | 200–150 | 0.60 |   | 150–100 | 0.46 |   | 40–20 | 0.00 |
|   | 100–50 | 6.33 |   | 90–50 | 0.00 |   | 20–0 | 0.05 |
| 7 | 500–400 | 0.63 | 15 | 300–200 | 0.01 | 21 | 40–20 | 0.01 |
|   | 300–250 | 1.35 |   | 150–100 | 0.41 |   | 20–0 | 0.00 |
|   | 200–150 | 0.39 |   | 80–60 | 0.04 |   |   |   |
|   | 100–50 | 6.46 |   | 40–20 | 0.00 |   |   |   |
| 9 | 700–600 | 0.04 | 16 | 80–60 | 0.26 | 22 | 30–20 | 0.01 |
|   | 400–300 | 0.57 |   | 60–40 | 0.00 |   | 20–10 | 0.00 |
|   | 200–140 | 4.15 |   | 40–20 | 0.00 |   | 10–0 | 0.00 |
|   | 100–50 | 7.02 |   | 20–0 | 0.00 |   |   |   |
| 10 | 500–400 | 0.13 | 17 | 120–100 | 1.57 | 23 | 30–0 | 0.00 |
|   | 300–250 | 0.88 |   | 80–60 | 0.05 |   |   |   |
|   | 200–150 | 3.49 |   | 40–30 | 0.00 |   |   |   |
|   | 100–50 | 1.64 |   | 20–10 | 0.05 |   |   |   |
| 11 | 400–300 | 0.25 | 18 | 160–120 | 0.35 | 26 | 30–20 | 0.02 |
|   | 250–200 | 1.53 |   | 100–80 | 0.32 |   | 20–0 | 0.00 |
|   | 150–100 | 2.52 |   | 60–40 | 0.03 |   |   |   |
|   | 50–25 | 0.03 |   | 20–10 | 0.00 |   |   |   |
| 12 | 130–100 | 0.33 |   |   |   |   |   |   |
|   | 100–50 | 0.37 |   |   |   |   |   |   |
|   | 50–25 | 0.02 |   |   |   |   |   |   |
|   | 25–0 | 0.09 |   |   |   |   |   |   |

*Macrocypridina poulseni*

| Station number | Depth (m) | Ind·m$^{-3}$ | Station number | Depth (m) | Ind·m$^{-3}$ | Station number | Depth (m) | Ind·m$^{-3}$ |
|---|---|---|---|---|---|---|---|---|
| 5 | 160–140 | 0.00 | 13 | 400–300 | 0.00 | 19 | 100–75 | 0.00 |
|   | 115–100 | 0.00 |   | 250–200 | 0.00 |   | 75–50 | 0.00 |
|   | 80–60 | 0.00 |   | 150–100 | 0.00 |   | 50–25 | 0.00 |
|   | 40–20 | 0.00 |   | 50–25 | 0.00 |   | 25–0 | 0.00 |
| 6 | 455–400 | 0.00 | 14 | 400–300 | 0.00 | 20 | 80–75 | 0.00 |
|   | 300–250 | 0.00 |   | 250–200 | 0.00 |   | 60–40 | 0.00 |
|   | 200–150 | 0.00 |   | 150–100 | 0.00 |   | 40–20 | 0.00 |
|   | 100–50 | 0.00 |   | 90–50 | 0.00 |   | 20–0 | 0.00 |
| 7 | 500–400 | 0.00 | 15 | 300–200 | 0.00 | 21 | 40–20 | 0.00 |
|   | 300–250 | 0.00 |   | 150–100 | 0.00 |   | 20–0 | 0.00 |
|   | 200–150 | 0.00 |   | 80–60 | 0.00 |   |   |   |
|   | 100–50 | 0.00 |   | 40–20 | 0.00 |   |   |   |
| 9 | 700–600 | 0.00 | 16 | 80–60 | 0.00 | 22 | 30–20 | 0.00 |
|   | 400–300 | 0.00 |   | 60–40 | 0.00 |   | 20–10 | 0.00 |
|   | 200–140 | 0.00 |   | 40–20 | 0.00 |   | 10–0 | 0.00 |
|   | 100–50 | 0.00 |   | 20–0 | 0.00 |   |   |   |
| 10 | 500–400 | 0.00 | 17 | 120–100 | 0.00 | 23 | 30–0 | 0.00 |
|   | 300–250 | 0.00 |   | 80–60 | 0.00 |   |   |   |
|   | 200–150 | 0.00 |   | 40–30 | 0.00 |   |   |   |
|   | 100–50 | 0.00 |   | 20–10 | 0.00 |   |   |   |
| 11 | 400–300 | 0.01 | 18 | 160–120 | 0.00 | 26 | 30–20 | 0.00 |
|   | 250–200 | 0.03 |   | 100–80 | 0.00 |   | 20–0 | 0.00 |
|   | 150–100 | 0.01 |   | 60–40 | 0.00 |   |   |   |
|   | 50–25 | 0.00 |   | 20–10 | 0.00 |   |   |   |
| 12 | 130–100 | 0.00 |   |   |   |   |   |   |
|   | 100–50 | 0.00 |   |   |   |   |   |   |
|   | 50–25 | 0.00 |   |   |   |   |   |   |
|   | 25–0 | 0.00 |   |   |   |   |   |   |

*Philomedes eugeniae*

| Station number | Depth (m) | Ind·m$^{-3}$ | Station number | Depth (m) | Ind·m$^{-3}$ | Station number | Depth (m) | Ind·m$^{-3}$ |
|---|---|---|---|---|---|---|---|---|
| 5 | 160–140 | 0.00 | 13 | 400–300 | 0.00 | 19 | 100–75 | 0.00 |
|   | 115–100 | 0.00 |    | 250–200 | 0.00 |    | 75–50 | 0.00 |
|   | 80–60 | 0.00 |    | 150–100 | 0.00 |    | 50–25 | 0.00 |
|   | 40–20 | 0.00 |    | 50–25 | 0.00 |    | 25–0 | 0.00 |
| 6 | 455–400 | 0.00 | 14 | 400–300 | 0.00 | 20 | 80–75 | 0.00 |
|   | 300–250 | 0.00 |    | 250–200 | 0.00 |    | 60–40 | 0.00 |
|   | 200–150 | 0.00 |    | 150–100 | 0.00 |    | 40–20 | 0.00 |
|   | 100–50 | 0.00 |    | 90–50 | 0.00 |    | 20–0 | 0.00 |
| 7 | 500–400 | 0.00 | 15 | 300–200 | 0.00 | 21 | 40–20 | 0.01 |
|   | 300–250 | 0.00 |    | 150–100 | 0.00 |    | 20–0 | 0.00 |
|   | 200–150 | 0.00 |    | 80–60 | 0.00 |    |   |   |
|   | 100–50 | 0.00 |    | 40–20 | 0.00 |    |   |   |
| 9 | 700–600 | 0.00 | 16 | 80–60 | 0.00 | 22 | 30–20 | 0.03 |
|   | 400–300 | 0.00 |    | 60–40 | 0.00 |    | 20–10 | 0.00 |
|   | 200–140 | 0.00 |    | 40–20 | 0.00 |    | 10–0 | 0.00 |
|   | 100–50 | 0.00 |    | 20–0 | 0.00 |    |   |   |
| 10 | 500–400 | 0.00 | 17 | 120–100 | 0.00 | 23 | 30–0 | 0.00 |
|   | 300–250 | 0.00 |    | 80–60 | 0.00 |    |   |   |
|   | 200–150 | 0.00 |    | 40–30 | 0.00 |    |   |   |
|   | 100–50 | 0.00 |    | 20–10 | 0.00 |    |   |   |
| 11 | 400–300 | 0.00 | 18 | 160–120 | 0.00 | 26 | 30–20 | 0.00 |
|   | 250–200 | 0.00 |    | 100–80 | 0.00 |    | 20–0 | 0.00 |
|   | 150–100 | 0.00 |    | 60–40 | 0.00 |    |   |   |
|   | 50–25 | 0.00 |    | 20–10 | 0.00 |    |   |   |
| 12 | 130–100 | 0.00 |    |   |   |    |   |   |
|   | 100–50 | 0.00 |    |   |   |    |   |   |
|   | 50–25 | 0.00 |    |   |   |    |   |   |
|   | 25–0 | 0.00 |    |   |   |    |   |   |

*Philomedes cubitum*

| Station number | Depth (m) | Ind·m$^{-3}$ | Station number | Depth (m) | Ind·m$^{-3}$ | Station number | Depth (m) | Ind·m$^{-3}$ |
|---|---|---|---|---|---|---|---|---|
| 5 | 160–140 | 0.00 | 13 | 400–300 | 0.00 | 19 | 100–75 | 0.00 |
|   | 115–100 | 0.00 |    | 250–200 | 0.00 |    | 75–50 | 0.00 |
|   | 80–60 | 0.00 |    | 150–100 | 0.00 |    | 50–25 | 0.00 |
|   | 40–20 | 0.00 |    | 50–25 | 0.00 |    | 25–0 | 0.00 |
| 6 | 455–400 | 0.00 | 14 | 400–300 | 0.00 | 20 | 80–75 | 0.00 |
|   | 300–250 | 0.00 |    | 250–200 | 0.00 |    | 60–40 | 0.00 |
|   | 200–150 | 0.00 |    | 150–100 | 0.00 |    | 40–20 | 0.00 |
|   | 100–50 | 0.00 |    | 90–50 | 0.00 |    | 20–0 | 0.00 |
| 7 | 500–400 | 0.00 | 15 | 300–200 | 0.00 | 21 | 40–20 | 0.01 |
|   | 300–250 | 0.00 |    | 150–100 | 0.00 |    | 20–0 | 0.00 |
|   | 200–150 | 0.00 |    | 80–60 | 0.00 |    |   |   |
|   | 100–50 | 0.00 |    | 40–20 | 0.00 |    |   |   |
| 9 | 700–600 | 0.00 | 16 | 80–60 | 0.00 | 22 | 30–20 | 0.00 |
|   | 400–300 | 0.00 |    | 60–40 | 0.00 |    | 20–10 | 0.00 |
|   | 200–140 | 0.00 |    | 40–20 | 0.00 |    | 10–0 | 0.00 |
|   | 100–50 | 0.00 |    | 20–0 | 0.00 |    |   |   |
| 10 | 500–400 | 0.00 | 17 | 120–100 | 0.00 | 23 | 30–0 | 0.00 |
|    | 300–250 | 0.00 |    | 80–60 | 0.00 |    |   |   |
|    | 200–150 | 0.00 |    | 40–30 | 0.00 |    |   |   |
|    | 100–50 | 0.00 |    | 20–10 | 0.00 |    |   |   |
| 11 | 400–300 | 0.00 | 18 | 160–120 | 0.00 | 26 | 30–20 | 0.00 |
|    | 250–200 | 0.00 |    | 100–80 | 0.00 |    | 20–0 | 0.00 |
|    | 150–100 | 0.00 |    | 60–40 | 0.00 |    |   |   |
|    | 50–25 | 0.00 |    | 20–10 | 0.00 |    |   |   |
| 12 | 130–100 | 0.00 |    |   |   |    |   |   |
|    | 100–50 | 0.00 |    |   |   |    |   |   |
|    | 50–25 | 0.00 |    |   |   |    |   |   |
|    | 25–0 | 0.00 |    |   |   |    |   |   |

*Paradoxostoma hypselum*

| Station number | Depth (m) | Ind·m$^{-3}$ | Station number | Depth (m) | Ind·m$^{-3}$ | Station number | Depth (m) | Ind·m$^{-3}$ |
|---|---|---|---|---|---|---|---|---|
| 5 | 160–140 | 0.00 | 13 | 400–300 | 0.00 | 19 | 100–75 | 0.00 |
|   | 115–100 | 0.00 |    | 250–200 | 0.00 |    | 75–50 | 0.00 |
|   | 80–60 | 0.00 |    | 150–100 | 0.00 |    | 50–25 | 0.00 |
|   | 40–20 | 0.00 |    | 50–25 | 0.00 |    | 25–0 | 0.00 |
| 6 | 455–400 | 0.00 | 14 | 400–300 | 0.00 | 20 | 80–75 | 0.04 |
|   | 300–250 | 0.00 |    | 250–200 | 0.00 |    | 60–40 | 0.00 |
|   | 200–150 | 0.00 |    | 150–100 | 0.00 |    | 40–20 | 0.00 |
|   | 100–50 | 0.00 |    | 90–50 | 0.00 |    | 20–0 | 0.00 |
| 7 | 500–400 | 0.00 | 15 | 300–200 | 0.00 | 21 | 40–20 | 0.06 |
|   | 300–250 | 0.00 |    | 150–100 | 0.00 |    | 20–0 | 0.00 |
|   | 200–150 | 0.00 |    | 80–60 | 0.00 |    |   |   |
|   | 100–50 | 0.00 |    | 40–20 | 0.00 |    |   |   |
| 9 | 700–600 | 0.00 | 16 | 80–60 | 0.00 | 22 | 30–20 | 0.36 |
|   | 400–300 | 0.00 |    | 60–40 | 0.00 |    | 20–10 | 0.28 |
|   | 200–140 | 0.00 |    | 40–20 | 0.00 |    | 10–0 | 0.07 |
|   | 100–50 | 0.00 |    | 20–0 | 0.00 |    |   |   |
| 10 | 500–400 | 0.00 | 17 | 120–100 | 0.00 | 23 | 30–0 | 0.00 |
|   | 300–250 | 0.00 |    | 80–60 | 0.00 |    |   |   |
|   | 200–150 | 0.00 |    | 40–30 | 0.00 |    |   |   |
|   | 100–50 | 0.00 |    | 20–10 | 0.00 |    |   |   |
| 11 | 400–300 | 0.00 | 18 | 160–120 | 0.00 | 26 | 30–20 | 0.00 |
|   | 250–200 | 0.00 |    | 100–80 | 0.00 |    | 20–0 | 0.00 |
|   | 150–100 | 0.00 |    | 60–40 | 0.00 |    |   |   |
|   | 50–25 | 0.03 |    | 20–10 | 0.00 |    |   |   |
| 12 | 130–100 | 0.00 |    |   |   |    |   |   |
|   | 100–50 | 0.00 |    |   |   |    |   |   |
|   | 50–25 | 0.00 |    |   |   |    |   |   |
|   | 25–0 | 0.00 |    |   |   |    |   |   |

*Paradoxostoma magellanicum*

| Station number | Depth (m) | Ind·m$^{-3}$ | Station number | Depth (m) | Ind·m$^{-3}$ | Station number | Depth (m) | Ind·m$^{-3}$ |
|---|---|---|---|---|---|---|---|---|
| 5 | 160–140 | 0.00 | 13 | 400–300 | 0.00 | 19 | 100–75 | 0.00 |
|   | 115–100 | 0.00 |    | 250–200 | 0.00 |    | 75–50 | 0.00 |
|   | 80–60 | 0.00 |    | 150–100 | 0.00 |    | 50–25 | 0.00 |
|   | 40–20 | 0.00 |    | 50–25 | 0.00 |    | 25–0 | 0.00 |
| 6 | 455–400 | 0.00 | 14 | 400–300 | 0.00 | 20 | 80–75 | 0.00 |
|   | 300–250 | 0.00 |    | 250–200 | 0.00 |    | 60–40 | 0.00 |
|   | 200–150 | 0.00 |    | 150–100 | 0.00 |    | 40–20 | 0.17 |
|   | 100–50 | 0.00 |    | 90–50 | 0.00 |    | 20–0 | 0.00 |
| 7 | 500–400 | 0.00 | 15 | 300–200 | 0.00 | 21 | 40–20 | 0.00 |
|   | 300–250 | 0.00 |    | 150–100 | 0.00 |    | 20–0 | 0.09 |
|   | 200–150 | 0.00 |    | 80–60 | 0.00 |    |  |  |
|   | 100–50 | 0.00 |    | 40–20 | 0.00 |    |  |  |
| 9 | 700–600 | 0.00 | 16 | 80–60 | 0.00 | 22 | 30–20 | 0.00 |
|   | 400–300 | 0.00 |    | 60–40 | 0.00 |    | 20–10 | 0.00 |
|   | 200–140 | 0.00 |    | 40–20 | 0.00 |    | 10–0 | 0.00 |
|   | 100–50 | 0.00 |    | 20–0 | 0.00 |    |  |  |
| 10 | 500–400 | 0.00 | 17 | 120–100 | 0.00 | 23 | 30–0 | 0.00 |
|    | 300–250 | 0.00 |    | 80–60 | 0.00 |    |  |  |
|    | 200–150 | 0.00 |    | 40–30 | 0.00 |    |  |  |
|    | 100–50 | 0.00 |    | 20–10 | 0.00 |    |  |  |
| 11 | 400–300 | 0.00 | 18 | 160–120 | 0.00 | 26 | 30–20 | 0.00 |
|    | 250–200 | 0.00 |    | 100–80 | 0.00 |    | 20–0 | 0.00 |
|    | 150–100 | 0.00 |    | 60–40 | 0.00 |    |  |  |
|    | 50–25 | 0.00 |    | 20–10 | 0.00 |    |  |  |
| 12 | 130–100 | 0.00 |  |  |  |  |  |  |
|    | 100–50 | 0.00 |  |  |  |  |  |  |
|    | 50–25 | 0.00 |  |  |  |  |  |  |
|    | 25–0 | 0.00 |  |  |  |  |  |  |

**Systematic Account**

Until 1965, the Phylum was subdivided only in genera. In that year, Tokioka (1965a,b) proposed a new and revolutionary classification based on morphological and phylogenetical (presently under discussion) characters. The phylum was divided into two classes: Archisagittoidea, which includes the presumed fossil chaetognath *Amiskwia sagittiformis*, and Sagittoidea which includes all the present-day chaetognaths. The latter class was in turn subdivided into two orders: Phragmophora which have transversal musculature (phragma) in the trunk, musculature which is absent in the Aphragmophora which have only primary and secondary muscle bands in the trunk and tail. The orders were subdivided in families and the hypertrophic genus *Sagitta* was subdivided into 11 genera. The old name *Sagitta* remains for a few species. All new taxa have new names. Since then, this classification has been further revised Salvini-Plaven (1986) and Bieri (1991a,b). At present, the number of described species of chaetognaths is almost 200, but many of these will probably pass into synonymy with others in the future.

Whereas Bieri (1991a,b) and many other specialists concord with the new classification, other authors such as Alvariño (1969), Pierrot-Bults (1974, 1976, 1979), Kapp (1980), Boltovskoy (1981), Furnestin and Ducret (1982) and Hagen (1985) use the traditional nomenclature. Kapp (1993) maintains that at present there is no valid phylogenetic basis to erect a new classification, and that the new method contradicts the rules of the International Code of Zoological Nomenclature. Authors such as Dinofrio (1973) follow the traditional nomenclature, but also report the new names proposed by Tokioka. This is also the criterion we have adopted in the present text.

**Old Classification**

Phylum CHAETOGNATHA Leuckart, 1854
Genus SAGITTA Quoy & Gaimard, 1827
7.1 *Sagitta decipiens* Fowler, 1905
7.2 *Sagitta gazellae* von Ritter-Zahony, 1909
7.3 *Sagitta maxima* Conant, 1896
7.4 *Sagitta tasmanica* (Thomson, 1947)

Genus EUKROHNIA von Ritter-Zahoney, 1909
7.5 *Eukrohnia hamata* Möbius, 1875

**New classification**

Phylum CHAETOGNATHA Leuckart, 1854
Class SAGITTOIDEA Claus & Grobben, 1905

Order APHRAGMOPHORA Tokioka, 1965
Family SAGITTIDAE Claus & Grobben, 1905

Genus *Mesosagitta* Tokioka, 1965
7.1 *Mesosagitta decipiens* Fowler, 1905
7.1 *Decipisagitta decipiens* Bieri, 1991
7.3 *Mesosagitta maxima* Conant, 1896
Genus *Flaccisagitta* Tokioka 1965
7.2 *Pseudosagitta gazellae* Bieri, 1991
Genus *Serratosagitta* Tokioka & Pathansali, 1963
7.4 *Serratosagitta tasmanica* (Thomson, 1947)

Family Eukrohniidae
Genus *Eukrohnia* von Ritter-Zahony, 1909
7.5 *Eukrohnia hamata* Möbius, 1875

## 7.1 *Sagitta decipiens* Fowler, 1905

*Sagitta decipiens* Fowler, 1905: p. 70, pl. 5 figs.
32–35 – Lea, 1955: p. 606, fig. 4 – Neto, 1961: p. 27,
figs. 22-24 – Vucetic, 1963: p. 67, fig. 7a – Ducret,
1968: p. 125, fig. 25 – Fagetti, 1968: p. 162 – Alvariño, 1969: p. 245, figs. 87a–c, 88a-k – Pierrot-
Bults, 1979: p. 137, figs. 1 a,d, 2–4 – Boltovskoy,
1981: p. 788, fig. 256, 8 – Bieri, 1991b: p. 221.
*Sagitta neodecipiens* Tokioka, 1959: p. 377, figs.
17–19, 20 – Ducret, 1962: p. 344, figs. 14–16, 18 –
Ducret, 1968: p. 125, fig. 25 – Gamulin, 1977:
p. 139 – Pierrot-Bults, 1979: p. 138, figs. 1–5.
*Mesosagitta decipiens* Tokioka, 1965a: p. 348 –
Bieri 1991a, p. 131 – Lutschinger, 1993: p. 28, fig.
13a,b.
*Decipisagitta decipiens* Bieri, 1991b: p. 223, fig. 3.

Body length –
    14.5–17.5 mm

**Description.** Body elongated, flaccid and transparent; slightly swollen in posterior third, just
before caudal septum. Intestinal diverticulae
present but not always easy to distinguish. Small
head bearing 6-7 hooks. Number of teeth variable: 6–8 triangular anterior teeth and 6–14 posterior teeth. Proportional length of tail with respect to total body length 24–28.5 %. Anterior
fins long and narrow with barely visible rays,
ending anteriorly behind ventral ganglion. Posterior fins with maximum width slightly behind
caudal septum, not reaching seminal vesicles.
Ovaries reaching posterior end or surpassing
anterior fins. When full, seminal vesicles flask-
like in shape and characteristic of the species,
with a glandular anterior extremity. In the Magellan specimens, ripe seminal vesicles were not
observed; their position was at an equal distance
between the posterior and caudal fins. The species may differ in aspect due mainly to the caudal
region, which can be notably opaque when
sperm are ripe.

### Remarks

There is much disaccord among systematicians
on the exact description of this species. In our
specimens, the number of hooks and teeth differ
from the number reported by other authors: 4–
10 anterior and 6–18 posterior teeth (Pierrot-
Bults 1979) and 8–10 anterior and 19–22 posteri-
or teeth (Alvariño 1969). The proportional
length of the tail with respect to the total body
length is similar to that reported by Fagetti
(22.7–27.2 %), Alvariño (25–29 %) and Pierrot-
Bults (25–32.7 %). Lutschinger (1993) draws
both anterior and posterior fins rayless in the
anterior region. The testicles are mature in specimens of 16 mm. In larger specimens, the characteristic club-shaped seminal receptacles are well
defined whereas the eggs are not yet mature. von
Ritter-Zahony (1911a) reports that the length
varies from 6–20 mm, whereas Pierrot-Bults
(1979) reports a maximum length of 13.5 mm.

### Distribution

The species is mesoplanktonic and is widely distributed in all oceans including the Antarctic,
where it has been defined by David (1958) as an
"exotic species". In the Straits of Magellan, it was
sampled almost exclusively in the western sector,
but one specimen was also found in Baia Inutil.
Pierrot-Bults (1979) maintains that *S. neodecipiens* is synonymous with *S. decipiens* Fowler.

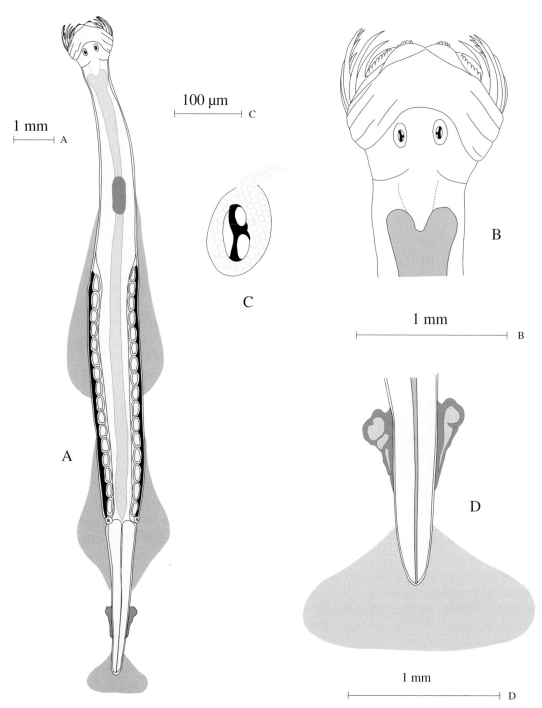

**Fig. 7.1.1A–D.** *Sagitta decipiens.* **A** Whole animal dorsal view; **B** detail of head; **C** eye; **D** detail of posterior end of tail and seminal vesicles. (Alvariño 1969)

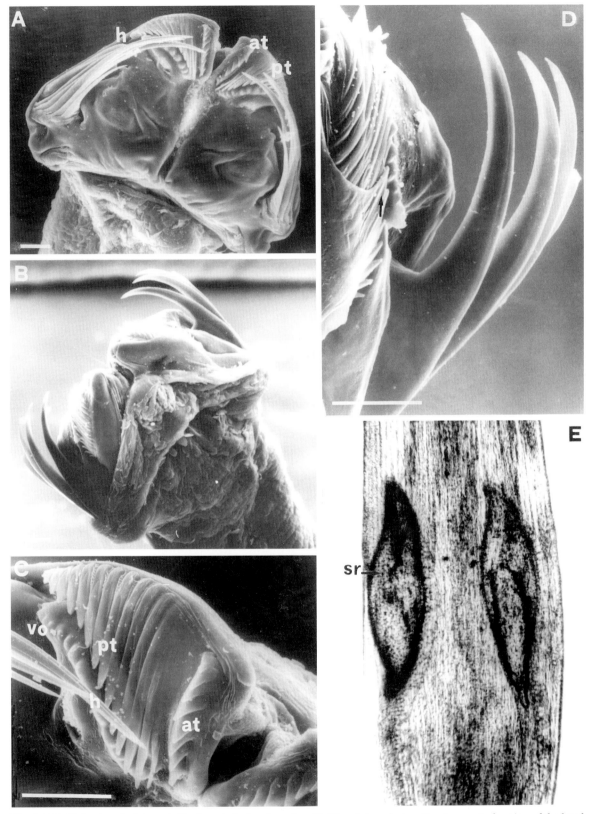

**Fig. 7.1.2A–E.** *Sagitta decipiens.* **A** Head ventral view (see also **C**); **B** head dorsal view; **C** anteroventral region of the head showing anterior teeth (*at*), posterior teeth (*pt*), vestibular organ (*vo*) and hooks (*h*); **D** detail of the posterior teeth, hooks or grasping spines; **E** light microscope image (x45) of seminal receptacles (*sr*). *Bars* **A,B,C,D** 100 µm

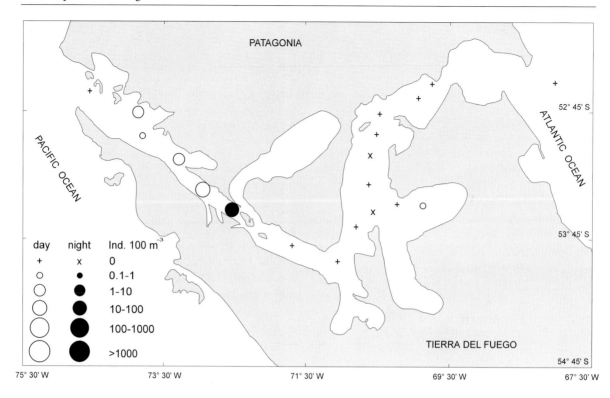

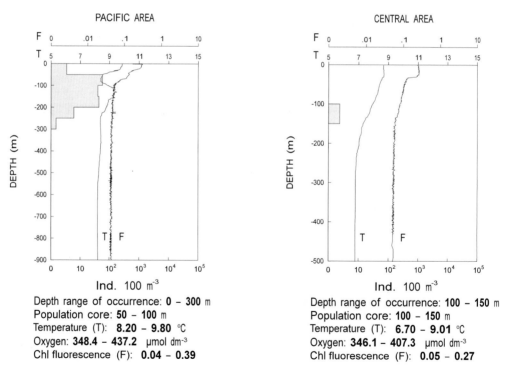

Depth range of occurrence: **0 – 300** m
Population core: **50 – 100** m
Temperature (T): **8.20 – 9.80** °C
Oxygen: **348.4 – 437.2** µmol dm⁻³
Chl fluorescence (F): **0.04 – 0.39**

Depth range of occurrence: **100 – 150** m
Population core: **100 – 150** m
Temperature (T): **6.70 – 9.01** °C
Oxygen: **346.1 – 407.3** µmol dm⁻³
Chl fluorescence (F): **0.05 – 0.27**

**Fig. 7.1.3.** *Sagitta decipiens.* Distribution in the Straits of Magellan in February–March 1991

## 7.2 *Sagitta gazellae* von Ritter-Zahony, 1909

*Sagitta gazellae* von Ritter-Zahony, 1909: pp. 787, 788 – von Ritter-Zahony 1911b: p. 10, figs. 4-7 – David, 1955: p. 246, figs. 1, 5a,b, 6b,c, 9, 10, 11a, b, 12a–d – Dinofrio, 1973: p. 26, pl. 1 fig. 3, pl. 2 fig. 3a,b – Alvariño, 1962: p. 27, fig. 5a,b – Alvariño, 1969: p. 158, figs. 48a-c, 49a-g – Boltovskoy, 1981: p. 788, fig. 256.
*Sagitta hexaptera* Fowler, 1907: p. 1.
*Sagitta lyra* Thomson, 1947: p. 11.
*Pseudosagitta gazellae* Bieri, 1991a: p. 128 – Lutschinger, 1993: p. 24, figs. 10a,b.

Body length –
  21–65 mm

**Description.** Body elongated, flaccid and transparent. Anterior fins much longer than posterior (almost double), commencing behind ventral ganglion. These are connected to the posterior fins by a narrow band. Rays are visible along external margin, especially in posterior half. Posterior fins subtriangular with rays along external border. Posterior extremity of fins almost reaching seminal vesicles. Corona ciliata pear-shaped and commencing anterior to eyes. Number of hooks varies from 5–10; number of anterior and posterior teeth varies from 5–10 and 3–10, respectively. Hooks and teeth more numerous in larger specimens.

### Remarks

The species has been reported to reach 105 mm in length in sub-Antarctic waters (David 1955) as compared to the maximum length of 65 mm recorded for Magellan specimens. This latter specimen had an ovary that was only 6 mm in length with eggs arranged along 8 rows. The longest ovary (23 mm) was still somewhat unripe and was observed in a specimen of 60 mm. The only specimen sampled in the eastern sector of the Straits, at the mouth of Baia Inutil, was 57 mm in length with an ovary of 22 mm, the eggs of which were still immature.

### Distribution

*S. gazellae* is an oceanic species widely distributed throughout the circumpolar Antarctic. It was originally considered only Antarctic in its distribution but, according to David (1955), maximum densities occur in sub-Antarctic regions, where it has been found to be 3 times as abundant as in the Antarctic. The northern limits of its distribution in the austral ocean correspond to the subtropical convergence, the position of which changes according to the Aghulas and SE Australian currents. It is probable that such changes also influence the northern limits of the distribution of *S. gazellae*. David (1955) records the presence of the species at the 35° parallel off the Argentine coast, whereas Kapp (1980) reports it somewhat more south at the 40° parallel. The species was also sampled in the Pacific Ocean, at the 50° parallel north of the Convergence, by Guglielmo et al. (1992). New records of *S. gazellae* are reported near New Zealand (North and South Islands) by Lutschinger (1993).

10 mm ⊢————⊣ A

100 µm ⊢———⊣ C

1 mm ⊢——⊣ B

1 mm ⊢——⊣ D

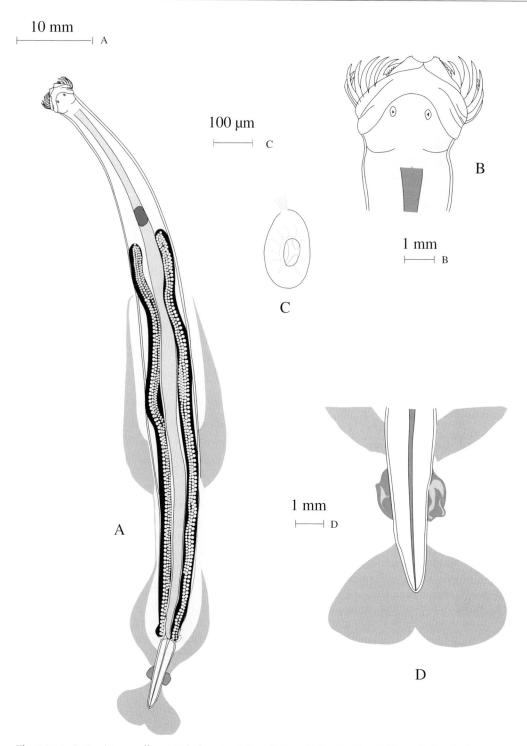

**Fig. 7.2.1A–D.** *Sagitta gazellae.* **A** Whole animal dorsal view; **B** detail of head; **C** eye; **D** detail of posterior end of tail and seminal vesicles. (Alvariño 1969)

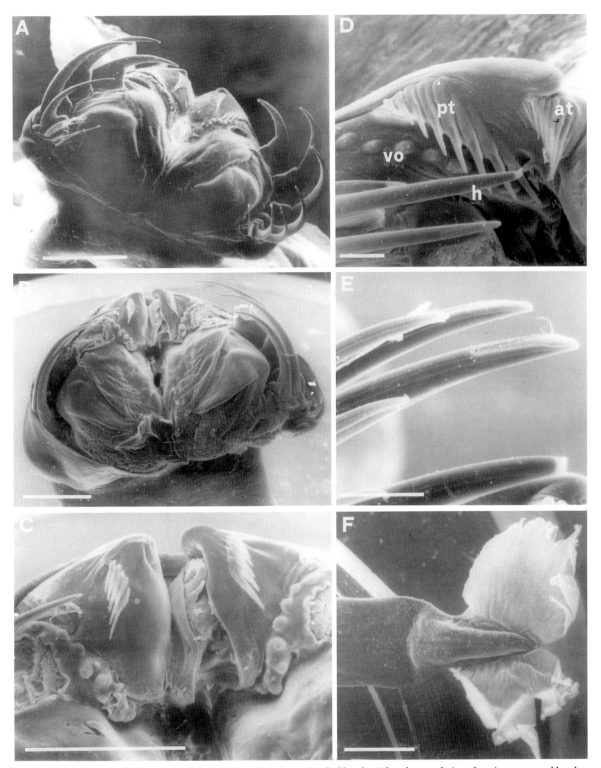

**Fig. 7.2.2A–F.** *Sagitta gazellae.* **A** Head ventral view showing extended hooks; **B** head ventral view showing retracted hooks; **C** posterior teeth and vestibular papillae clearly visible at the right side; **D** detail of the anterior teeth (*at*), posterior teeth (*pt*), vestibular papillae of the vestibular organ (*vo*) and hooks (*h*); **E** detail of poined hook extremities; **F** tail. *Bars* **A,B,F** 1 mm; **C** 500 µm; **D,E** 100 µm

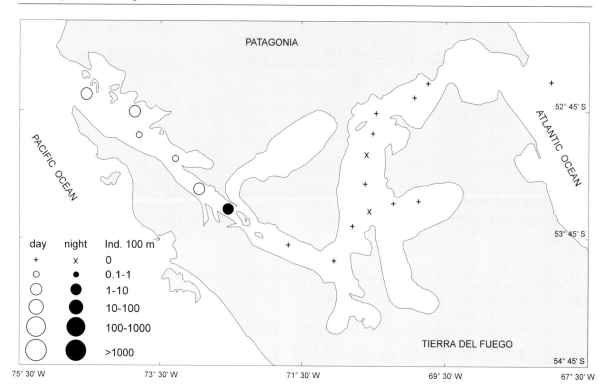

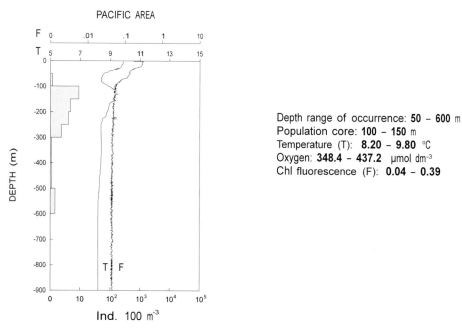

Depth range of occurrence: **50 – 600** m
Population core: **100 – 150** m
Temperature (T): **8.20 – 9.80** °C
Oxygen: **348.4 – 437.2** μmol dm[-3]
Chl fluorescence (F): **0.04 – 0.39**

**Fig. 7.2.3.** *Sagitta gazellae.* Distribution in the Straits of Magellan in February–March 1991

## 7.3 *Sagitta maxima* Conant, 1896

*Sagitta maxima* Conant, 1896: p. 84 – von Ritter-Zahony, 1910: p. 264, pl. 5 figs. 7, 8, 10 – von Ritter-Zahony, 1911b: p. 8, fig. 2 – David, 1958: p. 205, pl. 10 – Fagetti, 1968: p. 14, fig. 17a,b – Alvariño, 1962: p. 27, fig. 6a,b – Alvariño, 1969: p. 158, figs. 46a–d, 47a–f – Dinofrio, 1973: p. 42, fig. 6.
*Flaccisagitta maxima* Tokioka, 1965a: p. 351.
*Pseudosagitta maxima* Lutschinger, 1993: p. 24, fig. 11a,b.

Body length –
    juveniles: 15–43 mm

**Description.** Body elongated and flaccid, very similar in aspect to *S. gazellae,* with which it can be confused. Differences between the two species include the anterior fins, which originate at midlength of the ganglion in *S. maxima,* and the length of the tail, which is less than 15 % of the total length in *S. gazellae* and up to 19–25 % in *S. maxima.* The number of hooks (8–10) is greater in specimens 10–14 mm in length than in larger specimens, which have 5–6 hooks. The number of anterior and posterior teeth are 2–4 and 3–5, respectively. The anterior fins are united to the posterior fins. Rays visible posteriorly. Second pair of lateral fins subtriangular and bearing rays along external border. Ovary long and subtle with eggs arranged along 4–5 rows. In ripe specimens, ovary reaches or surpasses ventral ganglion.

**Remarks**

No mature specimens were collected in the present study. Arctic specimens are as long as 90 mm, whereas those in the Antarctic are 55 mm in length (David 1958).

**Distribution**

This is a cosmopolitan species present in the Arctic and Atlantic Oceans from 29°N to 74°S; in the Pacific it is sampled from 45°N to 46°S (Alvariño 1965). David (1958) reports its presence in the Pacific off Peru, where he observed a diminution in the number of hooks and teeth with increasing sexual maturity. Fagetti (1968) sampled it only in waters from the Peru current. Dinofrio (1973) recorded it in Drake Passage and the Weddel Sea; Guglielmo et al. (1992) collected it off South New Zealand. Lutschinger (1993) reports the species in New Zealand waters. In the Straits of Magellan, it was sampled only in the Pacific sector up to Cape Froward. *S. maxima* represented less than 1 % of all sampled chaetognaths.

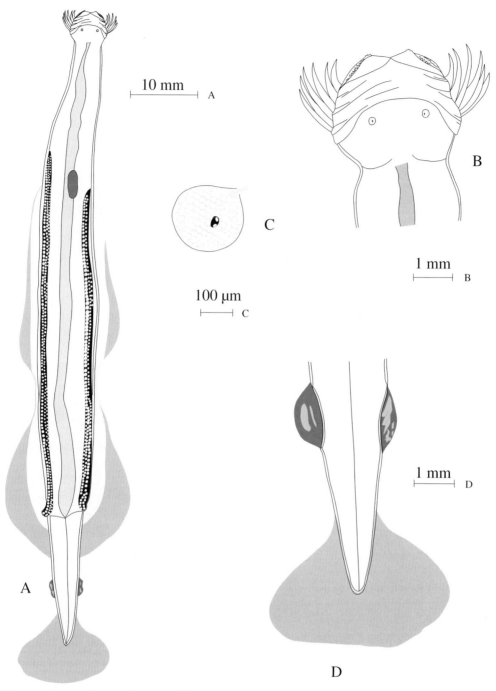

**Fig. 7.3.1A–D.** *Sagitta maxima.* **A** Whole animal dorsal view; **B** detail of head; **C** eye; **D** detail of posterior end of tail and seminal vesicles. (Alvariño 1969)

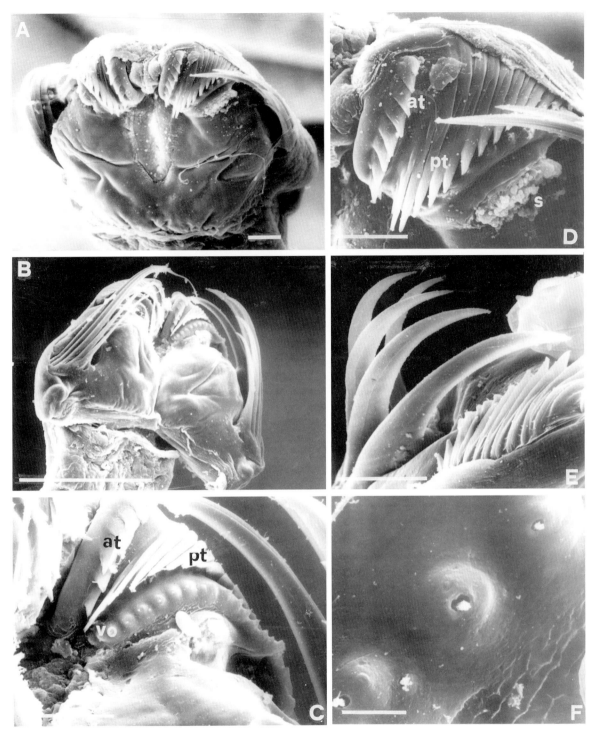

**Fig. 7.3.2 A–F.** *Sagitta maxima.* **A** Head anterior view; **B** ventral view; **C** head left side showing anterior teeth (*at*), posterior teeth (*pt*) and vestibular organ (*vo*); **D** anterior and posterior teeth, and secretion of vestibular papillae (*s*); **E** detail of hooks; **F** detail of vestibular papillae. *Bars* **A,C,D,E** 100 μm; **B** 500 μm; **F** 10 μm

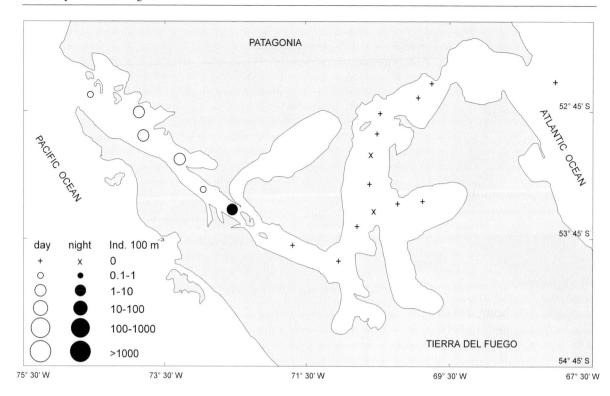

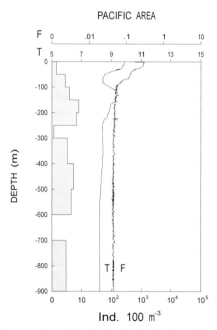

Depth range of occurrence: **0 – 900** m
Population core: **150 – 200** m
Temperature (T): **8.20 – 9.80** °C
Oxygen: **348.4 – 437.2** µmol dm⁻³
Chl fluorescence (F): **0.04 – 0.39**

**Fig. 7.3.3.** *Sagitta maxima.* Distribution in the Straits of Magellan in February–March 1991

## 7.4 *Sagitta tasmanica* Thomson, 1947

*Sagitta serrato-dentata* Krohn, 1853: p. 272, figs. 3, 4.
*Spadella serratodentata* Grassi, 1883: p. 19, pl. 1 figs. 8–10.
*Sagitta serratodentata* Strodtman, 1892: p. 39 – von Ritter-Zahony, 1911b: p. 22, figs. 21, 22 – Germain and Joubin, 1916: p. 41, pl. 3 figs. 12, 13, 19–21.
*Sagitta serratodentata tasmanica* Thomson, 1947: p. 15, figs 1–3 – Tokioka, 1959: p. 364, figs. 5, 7, 8.
*Sagitta tasmanica* Furnestin, 1957: p. 148, fig. 20, p. 151, fig. 54a–f – Fagetti, 1958a: p. 46, fig. 7 – Alvariño, 1961: p. 71, fig. 6 – Fagetti, 1968: p. 123, pl. 1 fig. a,b, pl. 3, figs. a–c – Ducret, 1968: p. 116, fig. 19a – Alvariño, 1969: pp. 188, 189, fig. 63a–d, fig. 64a–f – Pierrot-Bults, 1974: p. 220, figs. 4, 5 – Boltovskoy, 1981: p. 786, fig. 256 21a–d.
*Sagitta selkizki* Fagetti, 1958b: p. 125, pl. *1*, figs. a–d.
*Serratosagitta tasmanica* Lutschinger, 1993: p. 33, fig. 17.

Body length –
  10–24.5 mm

**Description.** Body elongated and rigid; total length/caudal length proportions varying from 21–33 %. Anterior fins beginning posterior to ventral ganglion. Posterior fins longer than anterior. Anterior portions of both anterior and posterior fins partially lacking rays. A narrow, barely visible band unites anterior and posterior fins. Posterior fins reaching seminal receptacles but not caudal fin. Mature and full receptacles appear as dark bulges. Mature ovaries variable in length and sometimes reaching beyond ventral ganglion in large specimens. Oocytes generally arranged in two rows but sometimes aligned in a single row. Eye pigment distributed in three branches. The number of hooks varies from 6–9 (more hooks observed in larger specimens). Hooks on head saw-edged at high magnification and appear as small mammellar protrusions (Fig. 7.4.2). Teeth vary from 4–6 anteriorly, and 8–15 posteriorly. Corona ciliata (ciliary loop) elliptical in shape, visible in very few specimens only after staining with methyl blue. Corona begins in front of eyes and extends onto neck and trunk. Distance from ventral ganglion twice the length of the ganglion. In large specimens (20–24 mm) the corona is 2.5–3 mm in length. There are no intestinal diverticulae.

**Remarks**

This was the most abundant species in the Straits of Magellan, representing 75 % of the chaetognaths sampled (Ghirardelli et al. 1995). The species shows considerable individual variability in size at sexual maturity, depending on the sampling area and depth. According to Pierrot-Bults (1974), the anterior portion of the anterior fins lacks rays, whereas Alvarino (1969) reports them as present. Fagetti (1968) found specimens in which the anterior fins lacked rays in southern Chilean regions, whereas they were present in northern specimens. In our specimens, the rays were partially lacking in both anterior and posterior fins. According to Fagetti (1968), the arrangement of the oocytes in a single or double row depends on the age of the specimens. Immature specimens have a double row of oocytes whereas mature specimens have only a single row. According to the same author, in southern Chilean waters individuals were already mature when 8–10 mm in length. Alvariño (1969) and

Fagetti (1968) reported that the length of mature specimens off the coast of Chile was 15–20 mm, somewhat smaller than those in our samples. The largest specimens were sampled at Sts. 14 and 20. Mainly immature specimens (5–7 mm) were sampled at western stations.

## Distribution

In the Atlantic Ocean, the species is distributed in two distinct geographical areas, from 35° to 65°N and from 35° and 65°S. The species has also been reported as sporadically present off the coast of Morocco, carried there by the cold Canary Current (Hernàndez and Jimenez 1993). In the southern Atlantic, the species is also transported northward by the cold Benguela Current, along a narrow area extending from the coast of Africa and that of Spain. In the Pacific and Indian Oceans, it shows a sub-Antarctic distribution delimited by the subtropical convergence (Alvariñ 1969). The species has also been sampled north of the Convergence. Lutschinger (1993) found it off South New Zealand. Biometric studies indicate that the northern and southern Atlantic morphotypes are similar. In the Pacific Ocean, the species is typically sub-Antarctic and is totally lacking in the Northern Hemisphere. In the present study, this was the only species present throughout the Straits of Magellan. It has also been sampled in the Beagle Channel (Fagetti Guaita 1959). Steinhaus (1900) first reports the presence in the Straits (Punta Arenas) of 8 young specimens classified as *S. serratodentata*.

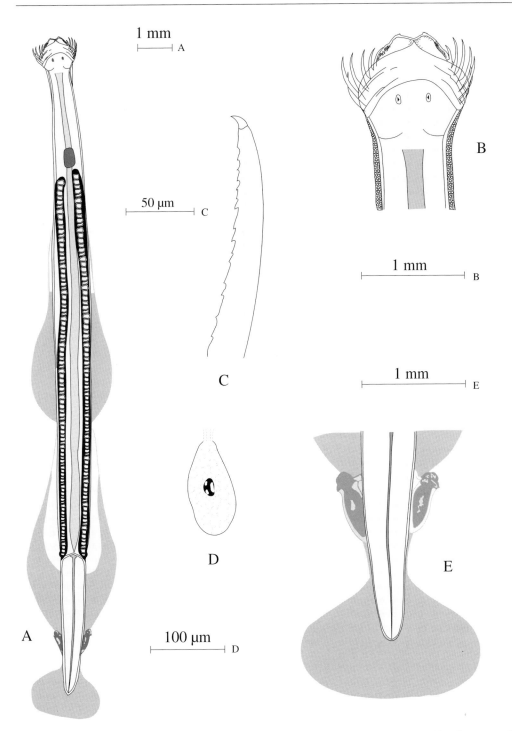

1 mm ⊢——⊣ A

50 μm ⊢——⊣ C

1 mm ⊢————⊣ B

1 mm ⊢————⊣ E

100 μm ⊢——⊣ D

A    B    C    D    E

**Fig. 7.4.1A–E.** *Sagitta tasmanica.* **A** Whole animal dorsal view; **B** detail of head; **C** serrated hook; **D** eye; **E** detail of posterior end of tail and seminal vesicles. (Alvariño 1969)

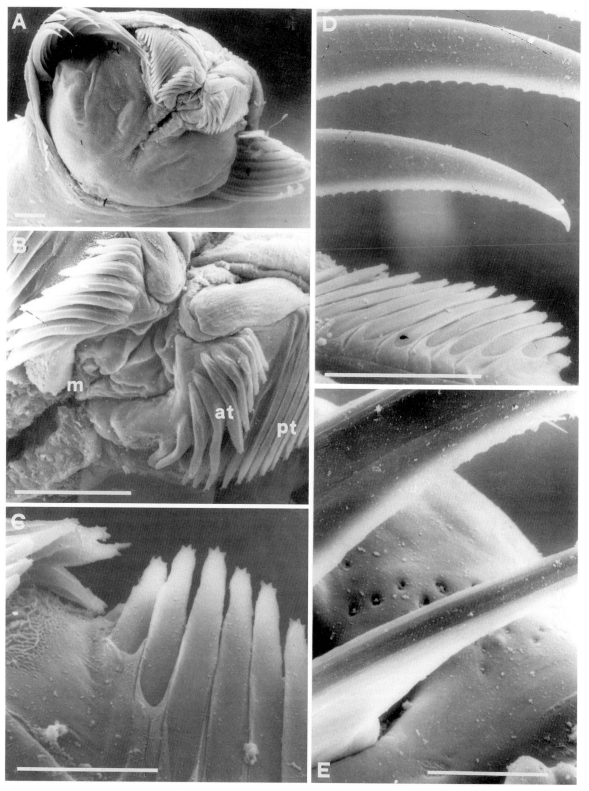

**Fig. 7.4.2A–E.** *Sagitta tasmanica.* **A** Head anteroventral view; **B** mouth (*m*) with anterior (*at*) and posterior teeth (*pt*); **C** detail of teeth with multicuspidate tips; **D** saw-shaped inner margin of hooks and posterior teeth; **E** transvestibular pores with cilia (bracket partially concealed by hooks). *Bars* **A,B,D** 100 μm; **C,E** 50 μm

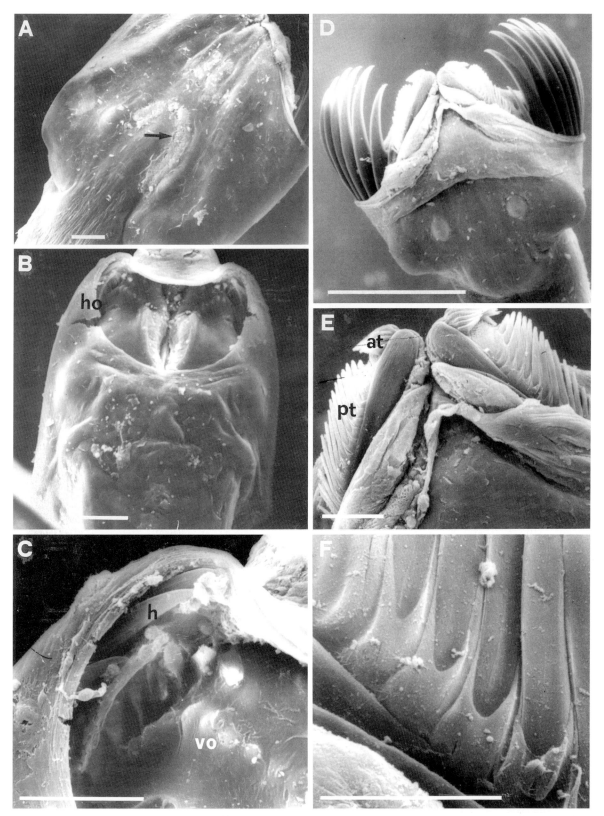

**Fig. 7.4.3A–F.** *Sagitta tasmanica.* **A** Head dorsal view with heavily damaged corona ciliata or ciliary loop (*arrowed*); **B** ventral view of head retracted within hood (*h*) or *praeputium*; **C** head retracted on hood, hooks (*h*) and vestibular organ (*vo*); **D** head dorsal view; eyes and extended hooks are visible; **E** ventral side, anterior and posterior teeth; **F** base of teeth. *Bars* **A,B,E** 100 μm; **C** 50 μm; **D,F** 500 μm

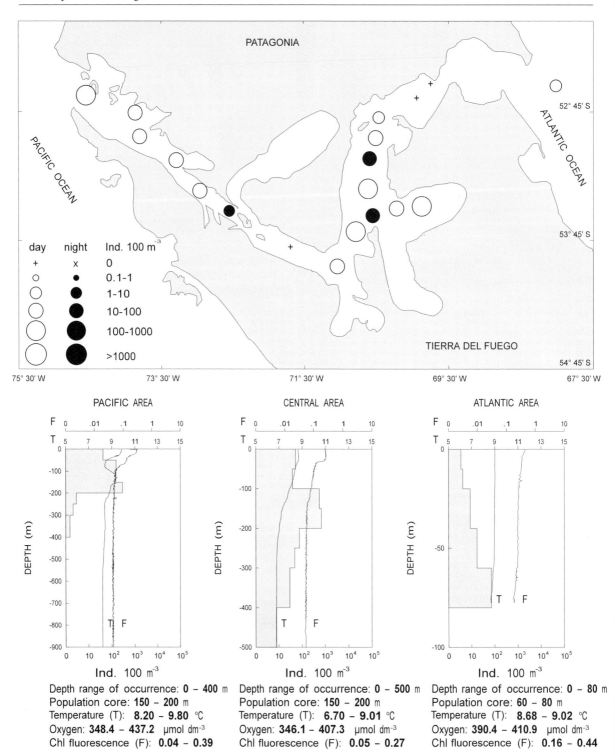

**Fig. 7.4.4.** *Sagitta tasmanica.* Distribution in the Straits of Magellan in February–March 1991

## 7.5 *Eukrohnia hamata* Möbius, 1875

*Eukrohnia hamata* Möbius, 1875: p. 158, pl. 3 figs. 13–16 – von Ritter-Zahony, 1910: p. 268, pl. 5 figs. 11a–b, 12 – Germain and Joubin, 1916: p. 58, pl. 1 figs. 2, 4, pl. 5 figs. 5–13, pl. 8 figs. 6,7 – Fagetti, 1958a: p. 67, fig. 16a–c – Neto, 1961: p. 46, figs. 38–40 – David, 1958: p. 206 – Ducret, 1965: p. 65, figs. 2, 3a, 4a – Dawson, 1968: p. 116, fig. 2a–d – Alvariño, 1969: p. 127, figs. 38a–d, 39a-i – Dinofrio, 1973: p. 18, pl. 1 fig. 5, pl. 2 fig. 5a,b, pl. 3 fig. 5 – pl. 5 fig. 1 – Boltovskoy, 1981: p. 786, fig. 256 – Lutschinger, 1993: p. 36, figs 20a–d.

Body length –
juveniles: 10-23 mm

**Description.** Individuals bear only 1 pair of delicate lateral fins. Maximum width of fins at level of caudal septum, with rays visible only along the posterior, external border. Anteriorly, fins originate at the level of the midventral ganglion. Posteriorly, fins end at about 2/3 distance separating distal extremity of tail from the caudal septum. Length of tail is 19–28 % of total length. Body generally flaccid, but large specimens may appear quite rigid. Commencing from the neck, the body progressively widens, reaching its maximum width at a distance from the caudal septum almost equal to the length of the tail. Primary body musculature well developed, conferring an opaque colour to the trunk. A thin transversal musculature is present. The head is small with an evident neck. There is no presence of alveolar tissue on the neck (i.e. collarette). There are 7–11 slightly recurved hooks (5-8 in the Magellan specimens), the conical extremities of which are bent at a right angle with respect to the hook's main axis. This is a distinctive character for the genus. The number of teeth varies from 6–26 (Alvariño 1969), with more teeth occurring in larger specimens. In our specimens, the number of teeth varied from 8–18, which more or less coincides with the number reported by von Ritter-Zahony (1911b) for Antarctic specimens and by Fagetti (1968) for specimens collected off the Chilean coast. According to Ducret (1965), more teeth occur in specimens off the African coast (23-25). Another characteristic feature of the species are the slightly protruding oval eyes which lack pigment and bear a small posterior appendix. Their surface consists of small lenticular structures similar to 'ommatidia', conferring to the species an aspect reminiscent of the compound eyes in arthropods. There are no esophageal diverticulae. Often, the intestine contains yellowish oily droplets. In mature specimens, the ovaries can be as long as 1/2 the trunk. The eggs are aligned in 4 rows. The seminal vesicles are oval and do not reach the caudal fin.

### Remarks

According to David (1958), the species shows a great variability in length. In the Southern Ocean, there are two distinct races that are more or less separated by the Antarctic Convergence. Larger individuals have been sampled in Antarctic waters, whereas subtropical specimens seem to belong to the smaller of the two races. Mature specimens reach 43 mm in length (Alvariño 1969; Boltovskoy 1981).

### Distribution

The species is epiplanktonic in Arctic and Antarctic regions. In tropical and equatorial areas, it has been sampled deeper, in intermediate depths, and in those greater than 900 m. During the *Siboga* expedition, it was sampled at 2000 m north of Australia (Fowler 1906). The species is considered bipolar in its distribution (David 1955; Alvariño 1969). Alvariño reports its presence at less than 100 m in Portugese waters, and in those off Guinea and in the Gulf of Biscay. She suggests that its presence in more surface waters may be due to upwelling of deeper waters. The species is present in the Atlantic, Indian and Pacific Oceans. It has also been reported for the Weddel Sea and Drake Passage (Dinofrio 1973), where very large specimens (up to 31.7 mm) have been sampled. It is also reported in the Gerlache Straits, Antarctic Peninsula. Ducret

(1968) reports it for the Atlantic coast of Tropical Africa at depths greater than 300 m, with maximum numbers occurring at even greater depths of 700–800 m. Fagetti (1968) sampled the species off the coast of Central Chile, but her specimens did not exceed 14.8 mm in length. von Ritter-Zahoni (1909) collected 2 very small specimens (9 and 12 mm) at the Pacific entrance to the Straits of Magellan during the *Gazelle* expedition. Guglielmo et al. (1992) sampled the species off South New Zealand. Lutschinger (1993) also reports the species along the New Zealand coast. In the present study, *E. hamata* was the second most abundant species (21.3 %). It was sampled mainly at Pacific stations and only 5 specimens were collected in the Atlantic sector.

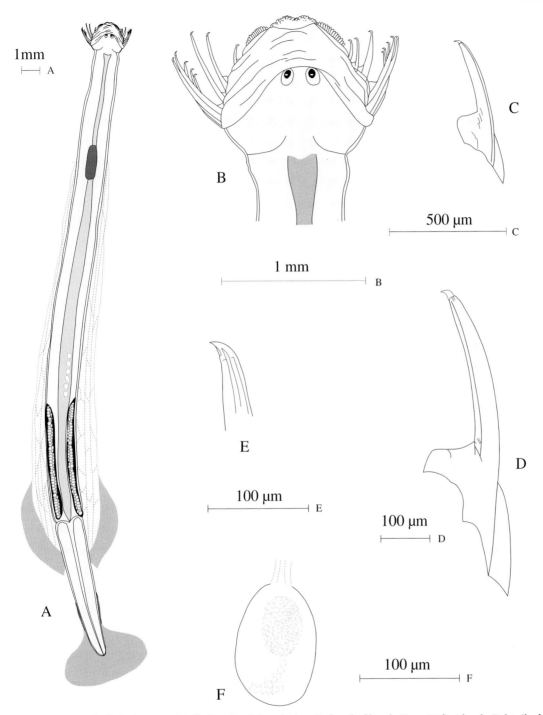

**Fig. 7.5.1A–F.** *Eukrohnia hamata.* **A** Whole animal dorsal view; **B** detail of head; **C** eye; **D** first hook; **E** detail of hook extremity; **F** second hook. (Alvariño 1969)

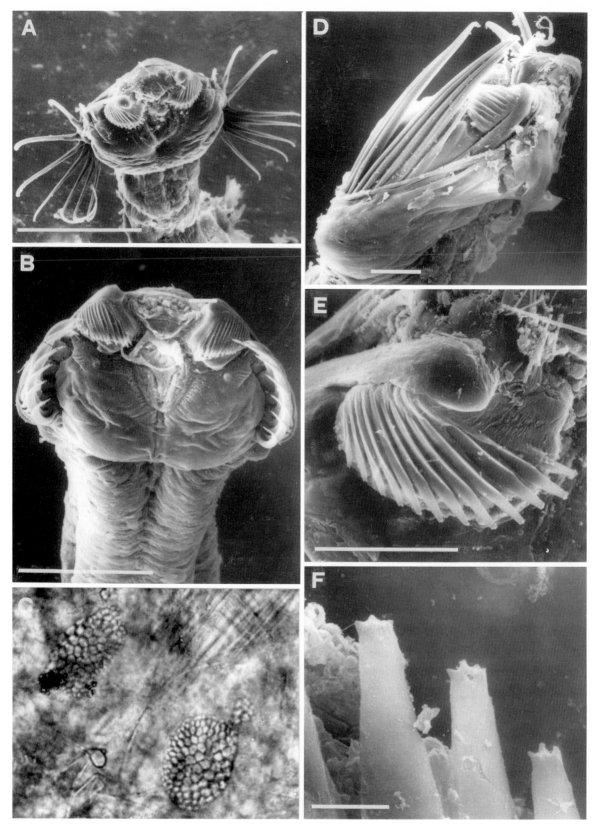

**Fig. 7.5.2A–F.** *Eukrohnia hamata.* **A** Antero ventral view showing extended hooks and posterior teeth (anterior teeth missing in *Eukrohnia*); **B** head ventral view; **C** light microscope image of eyes (x690); **D** head left side; hooks are partially enclosed in the hood; **E** ventral view of teeth; **F** jagged extremity of teeth. *Bars* **A,B** 500 μm; **D,E** 100 μm; **F** 10 μm

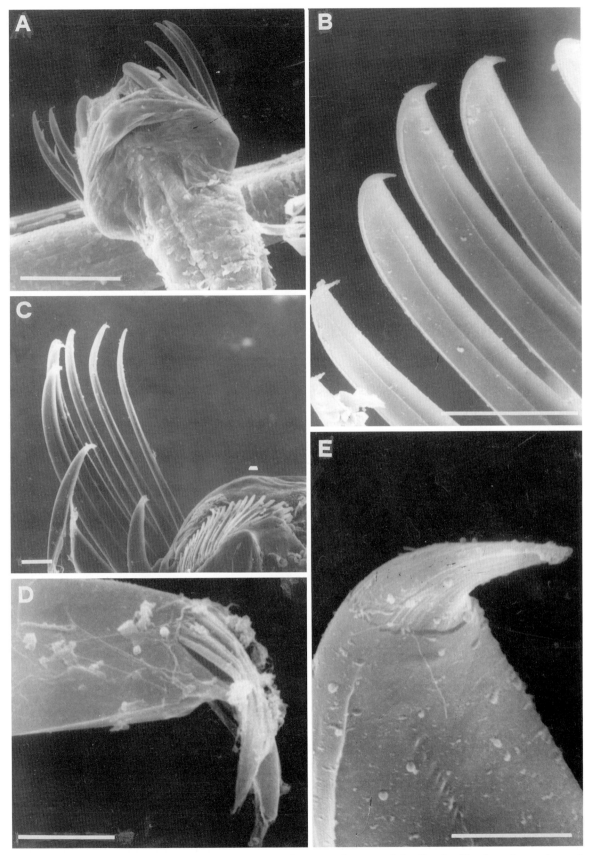

**Fig. 7.5.3A–F.** *Eukrohnia hamata.* **A** Head dorsal view; **B,E** characteristic recurved tip of hooks; **C** hooks and posterior teeth (right side, ventral view); **D** detail of hook showing the fibrous structure of the tip. *Bars* **A** 500 μm; **B,C** 100 μm; **D** 5 μm; **E** 10 μm

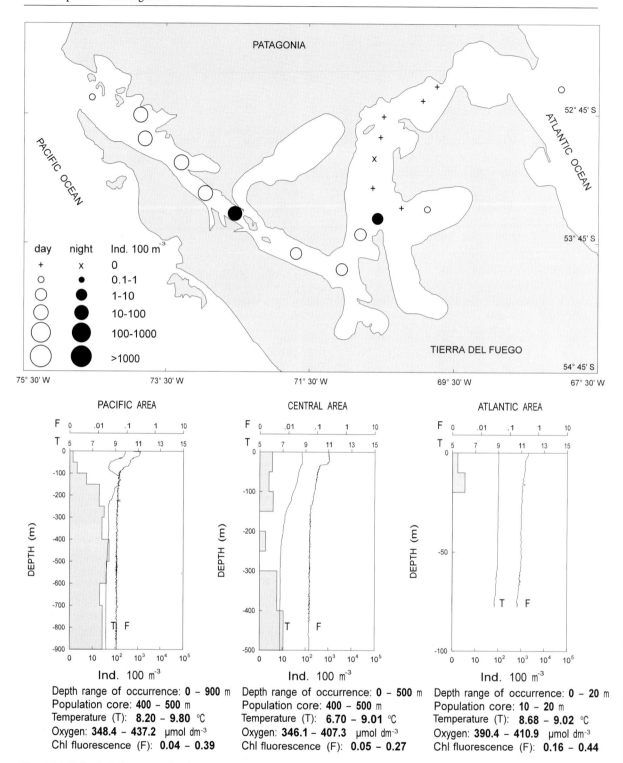

Fig. 7.5.4. *Eukrohnia hamata.* Distribution in the Straits of Magellan in February–March 1991

# References

Alvariño A (1961) Two new chaetognaths from the Pacific. Pac Sci 15: 67–77

Alvariño A (1962) Two new Pacific chaetognaths. Their distribution and relationship to allied species. Bull Scripps Inst Oceanogr Univ Calif 8: 1–50

Alvariño A (1965) Chaetognaths. Oceanogr Mar Biol Annu Rev 3: 115-194

Alvariño A (1969) Los quetognatos del Atlantico. Distribucion y notas esenciales de sistematica. Trab Inst Esp Oceanogr 37: 1–290

Bieri R (1991a) Systematics of the Chaetognatha. In: Bone Q, Kapp H, Pierrot-Bults AC (eds) The biology of chaetognaths. Oxford University Press, London, pp 122–136

Bieri R (1991b) Six new genera in the chaetognath family Sagittidae. Gulf Res Rep 8: 221–225

Boltovskoy D (1981) Chaetognatha. In: Boltovskoy D (ed) Atlas del zooplancton del Atlántico suboccidental y metodos de trabajo con el zooplancton marino. INDEP Mar del Plata, Argentina, pp 759–791

Claus C, Grobben K (1905) Leherbuch der Zoologie. NG Elwet´sche Verlagsbuch-handl, Marburg-Hessen

Conant FS (1896) Notes on the chaetognaths. Ann Mag Nat Hist 18: 201–214

David PM (1955) The distribution of *Sagitta gazellae* Ritter-Zahony. Discovery Rep 27: 235-278

David PM (1958) The distribution of the chaetognatha of the Southern Ocean. Discovery Rep 29: 199–228

Dawson JK (1968) Chaetognaths from the Arctic basin, including the description of a new species of *Heterokrohnia*. Bull South Calif Acad Sci 67: 112–124

Dinofrio EO (1973) Resultados planctologicos de la Campaña Oceantar I. 1 Quetognatos. Contrib Inst Antarct Argent 154: 1–62

Ducret F (1962) Chaetognathes des campagnes de l´*Ombango* dans la zone équatoriale africaine (1959-1960). Bull IFAN 24, Ser A, 2: 331–353

Ducret F (1965) Les espèces du genre *Eukrohnia* dans les eaux équatoriales et tropicales africaines. Cah ORSTOM 3(2): 63–78

Ducret F (1968) Chaetognathes des campagnes de l'*Ombango* dans les eaux équatoriales et tropicales africaines. Cah ORSTOM Sér Océanogr 6: 95–131

Fagetti Guaita E (1958a) Investigaciones sobre Quetognatos colectados, especialmente, frente a la costa central y norte de Chile. Rev Biol Mar 8: 25–81

Fagetti GE (1958b) Quetognato nuevo procedente del Archipelago de Juan Fernandez. Rev Biol Mar 8: 125–131

Fagetti Guaita E (1959) Quetognatos presentes en muestras Antárticas y Subantárcticas. Rev Biol Mar 9 (1,2,3): 251ª255

Fagetti GE (1968) Quetognatos de la expedicion *Marchile I* con observaciones acerca del posible valor de algunas especes como indicadoras de las masas de aqua frente a Chile. Rev Bio Mar 12: 85–171

Fowler GH (1905) Biscayan plankton collected during a cruise H.M.S. *Research* 1900, III. The Chaetognatha. Trans Linn Soc Lond (Zool) 10: 55–87

Fowler GH (1906) The Chaetognatha of the *Siboga* expedition with a discussion of the synonymy and distribution of the group. Siboga Exped Rep 21: 1–86

Fowler GH (1907) Chaetognatha, with a note on those collected by H.M.S. *Challenger* in subantarctic and antarctid waters. Nat Antarct Exped (1901-1904) 3:1–6

Furnestin ML (1957) Chaetognathes et zooplancton du secteur atlantique marocain. Rev Trav Inst Pêches Marit 21: 1–356

Furnestin ML, Ducret F (1982) Dents et organe vestibulaiore des chaetognathes au microscope électronique à balayage. Relations taxonomiques et biologiques. Rev Zool Afr 96: 138–173

Gamulin T (1977) Répartition des Chaetognathes en Mer Adriatique. Rapp Comm int Mer Médit 10 : 130–140

Germain L, Joubin L (1916) Chétognathes provenant des capagnes des yachts *Hirondelle* et Princesse Alice (1885–1910). Résult Camp Sci Prince Albert I. 49: 1–118

Ghirardelli E, Zagami G, Guglielmo L (1995) I chetognati dello Stretto di Magellano. Biol Mar Mediterr 2:537-539

Grassi GB (1883) I Chetognati. Fauna und flora des Golfes von Neapel Mon 5: 1–126

Guglielmo L, Costanzo G, Zagami G, Manganaro A, Arena G (1992) Zooplankton ecology in the Southern Ocean. Nat Sci Comm Antarct Ocean Camp Data Rep 1989-90 II: 301–468

Hagen W (1985) On distribution and population structure of Antarctic Chaetognata. Meeresforsch Rep Mar Res 30: 280–291

Hernandez F, Jimenez S (1993) Annual cycle of the chaetognaths of Los Cristianos (SW Tenerife, Canary Islands). In: Moreno I (ed) Proc II Int Workshop Chaetognatha. Palma, Universitat de les Illes Baleares, pp 121–127

Kapp H (1980) Result of the research cruises of FRV *Walter Herwig* in South America. Chaetognaths from the Patagonian Shelf in February 1971. Arch FishWiss 30: 125–135

Kapp H (1993) Some aspects of chaetognath systematics. In: Moreno I (ed) Proc II Int Worksh Chaetognatha. Palma, Universitat de les Illes Baleares, pp 37–43

Krohn A (1853) Nachträgliche Bemerkungen über den Bau Sagitta, nebst Beschreibung einiger neuen Arten. Arch Naturgesch 19: 26–281

Lea HE (1955) The chaetognaths of western Canadian coastal waters. J Fish Res Bd Can 12: 593–617

Leuckart R (1854) Bericht über die Leistungen der Naturgeschichte der niederen Thiere während der Jahre 1884-1853. Arch Naturgesch 20 (2): 334

Lutschinger S (1993) The marine fauna of New Zealand: Chaetognatha (arrow worms). N Z Oceanogr Inst Mem 101: 1–65

Möbius K (1875) Vermes in die Expedition zu physikalisch-chemischen und biologischen Untersuchungen der Nordsee im Sommer 1872. Wiss Meeresunters Kiel 2: 153–172

Neto FS (1961) Quetognatos dos mares de Angola. Trab Centro Biol Piscatoria 31: 9–60

Pierrot-Bults AC (1974) Taxonomy and zoogeography of certain members of the *Sagitta serratodentata* group (Chaetognatha). Bijdr Dierkd 44: 215-234

Pierrot-Bults AC (1976) Zoogeographic patterns in Chaetognatha and some other planktonic organisms. Bull Zool Mus Univ Amst 5: 59-72

Pierrot-Bults AC (1979) On the synonymy of *Sagitta decipiens* Fowler, 1905 and *Sagitta neodecipiens* Tokioka, 1959 and the validity of *Sagitta sibogae* Fowler, 1906. Bull Zool Mus Univ Amst 6: 137–143

Quoy J, Gaimard P (1827) Observations zoologiques faites a bord de l'Astrolabe en Mai 1826 dans le détroit de Gibraltar. Ann Sci Nat (Zool) 10: 5–239

von Ritter-Zahony R (1909) Die Chaetognaten der Gazelle-Expedition. Zool Anz 24: 787–793

von Ritter-Zahony R (1910) Die Chaetognaten. Fauna Arct, Jena 5: 249–288

von Ritter-Zahony R (1911a) Chaetognathi. Das Tierreich Lf 29: 1–36

von Ritter-Zahony R (1911b) Revision der Chätognathen. Dtsch Südpolar Exped Zool 13(5), Hf 1: 1–72

von Salvini-Plaven L (1986) Systematic notes on *Spadella* and on Chaetognatha in general. Z Zool Syst Evolutionsforsch 24: 122–128

Steinhaus O (1900) Chaetognathen. Hamb Magalhaenische Sammelreise 1(10): 658

Strodtmann S (1892) Die Systematik der Chetognathen und die geographische Verbreitung der einzelnen Arten im nordatlantischen Ocean. (Nach dem Material der Planktonexpedition 1889). Inaugural-Dissertation zur Erlangung der philosophischen Doctorwürde an der Christian Albrecht Universität zu Kiel Nicolaische Verlagshandlung, Berlin, pp 1–47

Thomson JM (1947) The Chaetognatha of Southeastern Australia. Bull Counc Sci Ind Res Div Fish Rep 14: 1–43

Tokioka T (1959) Observations on the taxonomy and distribution of chaetognaths in the North Pacific. Publ Seto Mar Lab 7: 349–356

Tokioka T (1965a) The taxonomical outline of chaetognaths. Publ Seto Mar Biol Lab 12: 335–357

Tokioka T (1965b) Supplementary notes on the systematics of Chaetognatha. Pubbl Seto Mar Biol Lab 13: 231–242

Tokioka T, Pathansali D (1963) Another new chaetognath from Malay waters, with a proposal of grouping some species of *Sagitta* into subgenera. Pubbl Seto Mar Biol Lab 11(1): 119–123

Vucetic T (1963) Zooplanktonic species as biological indicators of certain water masses. Hydrografski Godisnjak 1962: 73-80 (in Croatian)

## Stations and Sample Data for Chaetognaths During R/V *Cariboo* Cruise in the Straits of Magellan

*Sagitta decipiens*

| Station number | Depth (m) | Ind·m$^{-3}$ | Station number | Depth (m) | Ind·m$^{-3}$ | Station number | Depth (m) | Ind·m$^{-3}$ | Station number | Depth (m) | Ind·m$^{-3}$ | Station number | Depth (m) | Ind·m$^{-3}$ |
|---|---|---|---|---|---|---|---|---|---|---|---|---|---|---|
| 5 | 200–160 | 0.00 | 10 | 600–500 | 0.00 | 14 | 500–400 | 0.00 | 18 | 180–160 | 0.00 | 22 | 30–20 | 0.00 |
|  | 160–140 | 0.00 |  | 500–400 | 0.00 |  | 400–300 | 0.00 |  | 160–120 | 0.00 |  | 20–10 | 0.00 |
|  | 140–115 | 0.00 |  | 400–300 | 0.00 |  | 300–250 | 0.00 |  | 120–100 | 0.00 |  | 10–0 | 0.00 |
|  | 115–100 | 0.00 |  | 300–250 | 0.00 |  | 250–200 | 0.00 |  | 100–80 | 0.00 |  |  |  |
|  | 100–80 | 0.00 |  | 250–200 | 0.09 |  | 200–150 | 0.00 |  | 80–60 | 0.00 |  |  |  |
|  | 80–60 | 0.00 |  | 200–150 | 1.17 |  | 150–100 | 0.00 |  | 60–40 | 0.00 |  |  |  |
|  | 60–40 | 0.00 |  | 150–100 | 1.06 |  | 100–90 | 0.00 |  | 40–20 | 0.00 |  |  |  |
|  | 40–20 | 0.00 |  | 100–50 | 1.43 |  | 90–50 | 0.00 |  | 20–10 | 0.00 |  |  |  |
|  | 20–0 | 0.00 |  | 50–0 | 0.02 |  | 50–0 | 0.00 |  | 10–0 | 0.00 |  |  |  |
| 6 | 500–455 | 0.00 | 11 | 500–400 | 0.00 | 15 | 434–300 | 0.00 | 19 | 106–75 | 0.00 | 23 | 30–20 | 0.00 |
|  | 455–400 | 0.00 |  | 400–300 | 0.00 |  | 300–200 | 0.00 |  | 75–50 | 0.00 |  | 20–10 | 0.00 |
|  | 400–300 | 0.00 |  | 300–250 | 0.00 |  | 200–150 | 0.00 |  | 50–25 | 0.00 |  | 10–0 | 0.00 |
|  | 300–250 | 0.00 |  | 250–200 | 0.04 |  | 150–100 | 0.00 |  | 25–0 | 0.00 |  |  |  |
|  | 250–200 | 0.07 |  | 200–150 | 0.34 |  | 100–80 | 0.00 |  |  |  |  |  |  |
|  | 200–150 | 0.12 |  | 150–100 | 0.81 |  | 80–60 | 0.00 |  |  |  |  |  |  |
|  | 150–100 | 0.05 |  | 100–50 | 1.30 |  | 60–40 | 0.00 |  |  |  |  |  |  |
|  | 100–50 | 0.00 |  | 50–25 | 0.11 |  | 40–20 | 0.00 |  |  |  |  |  |  |
|  | 50–0 | 0.00 |  | 25–0 | 0.00 |  | 20–0 | 0.00 |  |  |  |  |  |  |
| 7 | 600–500 | 0.00 | 12 | 130–100 | 0.00 | 16 | 80–60 | 0.00 | 20 | 80–75 | 0.00 | 26 | 30–20 | 0.00 |
|  | 500–400 | 0.00 |  | 100–50 | 0.00 |  | 60–40 | 0.00 |  | 75–60 | 0.00 |  | 20–10 | 0.00 |
|  | 400–300 | 0.00 |  | 50–25 | 0.00 |  | 40–20 | 0.00 |  | 60–40 | 0.00 |  | 10–0 | 0.00 |
|  | 300–250 | 0.02 |  | 25–0 | 0.00 |  | 20–0 | 0.00 |  | 40–20 | 0.00 |  |  |  |
|  | 250–200 | 0.01 |  |  |  |  |  |  |  | 20–0 | 0.00 |  |  |  |
|  | 200–150 | 0.02 |  |  |  |  |  |  |  |  |  |  |  |  |
|  | 150–100 | 0.00 |  |  |  |  |  |  |  |  |  |  |  |  |
|  | 100–50 | 0.00 |  |  |  |  |  |  |  |  |  |  |  |  |
|  | 50–0 | 0.00 |  |  |  |  |  |  |  |  |  |  |  |  |
| 9 | 900–700 | 0.00 | 13 | 450–400 | 0.00 | 17 | 140–120 | 0.00 | 21 | 40–20 | 0.00 |  |  |  |
|  | 700–600 | 0.00 |  | 400–300 | 0.00 |  | 120–100 | 0.03 |  | 20–0 | 0.00 |  |  |  |
|  | 600–400 | 0.00 |  | 300–250 | 0.00 |  | 100–80 | 0.00 |  |  |  |  |  |  |
|  | 400–300 | 0.00 |  | 250–200 | 0.00 |  | 80–60 | 0.00 |  |  |  |  |  |  |
|  | 300–200 | 0.00 |  | 200–150 | 0.00 |  | 60–40 | 0.00 |  |  |  |  |  |  |
|  | 200–140 | 0.48 |  | 150–100 | 0.00 |  | 40–30 | 0.00 |  |  |  |  |  |  |
|  | 140–100 | 0.05 |  | 100–50 | 0.00 |  | 30–20 | 0.00 |  |  |  |  |  |  |
|  | 100–50 | 0.10 |  | 50–25 | 0.00 |  | 20–10 | 0.00 |  |  |  |  |  |  |
|  | 50–0 | 0.02 |  | 25–0 | 0.00 |  | 10–0 | 0.00 |  |  |  |  |  |  |

*Sagitta gazellae*

| Station number | Depth (m) | Ind·m$^{-3}$ | Station number | Depth (m) | Ind·m$^{-3}$ | Station number | Depth (m) | Ind·m$^{-3}$ | Station number | Depth (m) | Ind·m$^{-3}$ | Station number | Depth (m) | Ind·m$^{-3}$ |
|---|---|---|---|---|---|---|---|---|---|---|---|---|---|---|
| 5 | 200–160 | 0.08 | 10 | 600–500 | 0.01 | 14 | 500–400 | 0.00 | 18 | 180–160 | 0.00 | 22 | 30–20 | 0.00 |
|  | 160–140 | 0.07 |  | 500–400 | 0.01 |  | 400–300 | 0.00 |  | 160–120 | 0.00 |  | 20–10 | 0.00 |
|  | 140–115 | 0.02 |  | 400–300 | 0.00 |  | 300–250 | 0.00 |  | 120–100 | 0.00 |  | 10–0 | 0.00 |
|  | 115–100 | 0.00 |  | 300–250 | 0.01 |  | 250–200 | 0.00 |  | 100–80 | 0.00 |  |  |  |
|  | 100–80 | 0.00 |  | 250–200 | 0.00 |  | 200–150 | 0.00 |  | 80–60 | 0.00 |  |  |  |
|  | 80–60 | 0.00 |  | 200–150 | 0.08 |  | 150–100 | 0.00 |  | 60–40 | 0.00 |  |  |  |
|  | 60–40 | 0.00 |  | 150–100 | 0.09 |  | 100–90 | 0.00 |  | 40–20 | 0.00 |  |  |  |
|  | 40–20 | 0.00 |  | 100–50 | 0.00 |  | 90–50 | 0.00 |  | 20–10 | 0.00 |  |  |  |
|  | 20–0 | 0.00 |  | 50–0 | 0.00 |  | 50–0 | 0.00 |  | 10–0 | 0.00 |  |  |  |
| 6 | 500–455 | 0.00 | 11 | 500–400 | 0.00 | 15 | 434–300 | 0.00 | 19 | 106–75 | 0.00 | 23 | 30–20 | 0.00 |
|  | 455–400 | 0.00 |  | 400–300 | 0.01 |  | 300–200 | 0.00 |  | 75–50 | 0.00 |  | 20–10 | 0.00 |
|  | 400–300 | 0.00 |  | 300–250 | 0.02 |  | 200–150 | 0.00 |  | 50–25 | 0.00 |  | 10–0 | 0.00 |
|  | 300–250 | 0.03 |  | 250–200 | 0.03 |  | 150–100 | 0.00 |  | 25–0 | 0.00 |  |  |  |
|  | 250–200 | 0.09 |  | 200–150 | 0.04 |  | 100–80 | 0.00 |  |  |  |  |  |  |
|  | 200–150 | 0.00 |  | 150–100 | 0.17 |  | 80–60 | 0.00 |  |  |  |  |  |  |
|  | 150–100 | 0.10 |  | 100–50 | 0.00 |  | 60–40 | 0.00 |  |  |  |  |  |  |
|  | 100–50 | 0.01 |  | 50–25 | 0.00 |  | 40–20 | 0.00 |  |  |  |  |  |  |
|  | 50–0 | 0.00 |  | 25–0 | 0.00 |  | 20–0 | 0.00 |  |  |  |  |  |  |
| 7 | 600–500 | 0.00 | 12 | 130–100 | 0.00 | 16 | 80–60 | 0.00 | 20 | 80–75 | 0.00 | 26 | 30–20 | 0.00 |
|  | 500–400 | 0.00 |  | 100–50 | 0.00 |  | 60–40 | 0.00 |  | 75–60 | 0.00 |  | 20–10 | 0.00 |
|  | 400–300 | 0.00 |  | 50–25 | 0.00 |  | 40–20 | 0.00 |  | 60–40 | 0.00 |  | 10–0 | 0.00 |
|  | 300–250 | 0.00 |  | 25–0 | 0.00 |  | 20–0 | 0.00 |  | 40–20 | 0.00 |  |  |  |
|  | 250–200 | 0.01 |  |  |  |  |  |  |  | 20–0 | 0.00 |  |  |  |
|  | 200–150 | 0.04 |  |  |  |  |  |  |  |  |  |  |  |  |
|  | 150–100 | 0.00 |  |  |  |  |  |  |  |  |  |  |  |  |
|  | 100–50 | 0.00 |  |  |  |  |  |  |  |  |  |  |  |  |
|  | 50–0 | 0.00 |  |  |  |  |  |  |  |  |  |  |  |  |
| 9 | 900–700 | 0.00 | 13 | 450–400 | 0.00 | 17 | 140–120 | 0.00 | 21 | 40–20 | 0.00 |  |  |  |
|  | 700–600 | 0.00 |  | 400–300 | 0.00 |  | 120–100 | 0.00 |  | 20–0 | 0.00 |  |  |  |
|  | 600–400 | 0.00 |  | 300–250 | 0.00 |  | 100–80 | 0.00 |  |  |  |  |  |  |
|  | 400–300 | 0.00 |  | 250–200 | 0.00 |  | 80–60 | 0.00 |  |  |  |  |  |  |
|  | 300–200 | 0.02 |  | 200–150 | 0.00 |  | 60–40 | 0.00 |  |  |  |  |  |  |
|  | 200–140 | 0.00 |  | 150–100 | 0.00 |  | 40–30 | 0.00 |  |  |  |  |  |  |
|  | 140–100 | 0.09 |  | 100–50 | 0.00 |  | 30–20 | 0.00 |  |  |  |  |  |  |
|  | 100–50 | 0.00 |  | 50–25 | 0.00 |  | 20–10 | 0.00 |  |  |  |  |  |  |
|  | 50–0 | 0.00 |  | 25–0 | 0.00 |  | 10–0 | 0.00 |  |  |  |  |  |  |

*Sagitta maxima*

| Station number | Depth (m) | Ind·m$^{-3}$ | Station number | Depth (m) | Ind·m$^{-3}$ | Station number | Depth (m) | Ind·m$^{-3}$ | Station number | Depth (m) | Ind·m$^{-3}$ | Station number | Depth (m) | Ind·m$^{-3}$ |
|---|---|---|---|---|---|---|---|---|---|---|---|---|---|---|
| 5 | 200–160 | 0.02 | 10 | 600–500 | 0.00 | 14 | 500–400 | 0.00 | 18 | 180–160 | 0.00 | 22 | 30–20 | 0.00 |
|  | 160–140 | 0.02 |  | 500–400 | 0.02 |  | 400–300 | 0.00 |  | 160–120 | 0.00 |  | 20–10 | 0.00 |
|  | 140–115 | 0.00 |  | 400–300 | 0.00 |  | 300–250 | 0.00 |  | 120–100 | 0.00 |  | 10–0 | 0.00 |
|  | 115–100 | 0.00 |  | 300–250 | 0.00 |  | 250–200 | 0.00 |  | 100–80 | 0.00 |  |  |  |
|  | 100–80 | 0.00 |  | 250–200 | 0.01 |  | 200–150 | 0.00 |  | 80–60 | 0.00 |  |  |  |
|  | 80–60 | 0.00 |  | 200–150 | 0.00 |  | 150–100 | 0.00 |  | 60–40 | 0.00 |  |  |  |
|  | 60–40 | 0.00 |  | 150–100 | 0.00 |  | 100–90 | 0.00 |  | 40–20 | 0.00 |  |  |  |
|  | 40–20 | 0.00 |  | 100–50 | 0.00 |  | 90–50 | 0.00 |  | 20–10 | 0.00 |  |  |  |
|  | 20–0 | 0.00 |  | 50–0 | 0.00 |  | 50–0 | 0.00 |  | 10–0 | 0.00 |  |  |  |
| 6 | 500–455 | 0.12 | 11 | 500–400 | 0.01 | 15 | 434–300 | 0.00 | 19 | 106–75 | 0.00 | 23 | 30–20 | 0.00 |
|  | 455–400 | 0.04 |  | 400–300 | 0.01 |  | 300–200 | 0.00 |  | 75–50 | 0.00 |  | 20–10 | 0.00 |
|  | 400–300 | 0.02 |  | 300–250 | 0.01 |  | 200–150 | 0.00 |  | 50–25 | 0.00 |  | 10–0 | 0.00 |
|  | 300–250 | 0.00 |  | 250–200 | 0.02 |  | 150–100 | 0.00 |  | 25–0 | 0.00 |  |  |  |
|  | 250–200 | 0.09 |  | 200–150 | 0.01 |  | 100–80 | 0.00 |  |  |  |  |  |  |
|  | 200–150 | 0.30 |  | 150–100 | 0.04 |  | 80–60 | 0.00 |  |  |  |  |  |  |
|  | 150–100 | 0.00 |  | 100–50 | 0.01 |  | 60–40 | 0.00 |  |  |  |  |  |  |
|  | 100–50 | 0.00 |  | 50–25 | 0.04 |  | 40–20 | 0.00 |  |  |  |  |  |  |
|  | 50–0 | 0.00 |  | 25–0 | 0.00 |  | 20–0 | 0.00 |  |  |  |  |  |  |
| 7 | 600–500 | 0.07 | 12 | 130–100 | 0.00 | 16 | 80–60 | 0.00 | 20 | 80–75 | 0.00 | 26 | 30–20 | 0.00 |
|  | 500–400 | 0.03 |  | 100–50 | 0.00 |  | 60–40 | 0.00 |  | 75–60 | 0.00 |  | 20–10 | 0.00 |
|  | 400–300 | 0.04 |  | 50–25 | 0.00 |  | 40–20 | 0.00 |  | 60–40 | 0.00 |  | 10–0 | 0.00 |
|  | 300–250 | 0.00 |  | 25–0 | 0.00 |  | 20–0 | 0.00 |  | 40–20 | 0.00 |  |  |  |
|  | 250–200 | 0.12 |  |  |  |  |  |  |  | 20–0 | 0.00 |  |  |  |
|  | 200–150 | 0.05 |  |  |  |  |  |  |  |  |  |  |  |  |
|  | 150–100 | 0.08 |  |  |  |  |  |  |  |  |  |  |  |  |
|  | 100–50 | 0.04 |  |  |  |  |  |  |  |  |  |  |  |  |
|  | 50–0 | 0.00 |  |  |  |  |  |  |  |  |  |  |  |  |
| 9 | 900–700 | 0.02 | 13 | 450–400 | 0.00 | 17 | 140–120 | 0.00 | 21 | 40–20 | 0.00 |  |  |  |
|  | 700–600 | 0.00 |  | 400–300 | 0.00 |  | 120–100 | 0.00 |  | 20–0 | 0.00 |  |  |  |
|  | 600–400 | 0.01 |  | 300–250 | 0.00 |  | 100–80 | 0.00 |  |  |  |  |  |  |
|  | 400–300 | 0.06 |  | 250–200 | 0.00 |  | 80–60 | 0.00 |  |  |  |  |  |  |
|  | 300–200 | 0.02 |  | 200–150 | 0.00 |  | 60–40 | 0.00 |  |  |  |  |  |  |
|  | 200–140 | 0.05 |  | 150–100 | 0.00 |  | 40–30 | 0.00 |  |  |  |  |  |  |
|  | 140–100 | 0.09 |  | 100–50 | 0.00 |  | 30–20 | 0.00 |  |  |  |  |  |  |
|  | 100–50 | 0.08 |  | 50–25 | 0.00 |  | 20–10 | 0.00 |  |  |  |  |  |  |
|  | 50–0 | 0.00 |  | 25–0 | 0.00 |  | 10–0 | 0.00 |  |  |  |  |  |  |

*Sagitta tasmanica*

| Station number | Depth (m) | Ind·m⁻³ | Station number | Depth (m) | Ind·m⁻³ | Station number | Depth (m) | Ind·m⁻³ | Station number | Depth (m) | Ind·m⁻³ | Station number | Depth (m) | Ind·m⁻³ |
|---|---|---|---|---|---|---|---|---|---|---|---|---|---|---|
| 5 | 200–160 | 13.39 | 10 | 600–500 | 0.00 | 14 | 500–400 | 0.09 | 18 | 180–160 | 18.52 | 22 | 30–20 | 0.00 |
|  | 160–140 | 5.71 |  | 500–400 | 0.00 |  | 400–300 | 0.51 |  | 160–120 | 5.21 |  | 20–10 | 0.00 |
|  | 140–115 | 0.66 |  | 400–300 | 0.00 |  | 300–250 | 0.84 |  | 120–100 | 0.08 |  | 10–0 | 0.00 |
|  | 115–100 | 3.75 |  | 300–250 | 0.01 |  | 250–200 | 1.23 |  | 100–80 | 0.21 |  |  |  |
|  | 100–80 | 0.39 |  | 250–200 | 0.01 |  | 200–150 | 6.00 |  | 80–60 | 0.00 |  |  |  |
|  | 80–60 | 3.24 |  | 200–150 | 0.28 |  | 150–100 | 2.22 |  | 60–40 | 0.06 |  |  |  |
|  | 60–40 | 0.22 |  | 150–100 | 0.37 |  | 100–90 | 0.13 |  | 40–20 | 0.03 |  |  |  |
|  | 40–20 | 0.21 |  | 100–50 | 4.43 |  | 90–50 | 0.10 |  | 20–10 | 0.08 |  |  |  |
|  | 20–0 | 0.36 |  | 50–0 | 0.03 |  | 50–0 | 0.10 |  | 10–0 | 0.58 |  |  |  |
| 6 | 500–455 | 0.00 | 11 | 500–400 | 0.00 | 15 | 434–300 | 0.24 | 19 | 106–75 | 1.51 | 23 | 30–20 | 0.00 |
|  | 455–400 | 0.00 |  | 400–300 | 0.02 |  | 300–200 | 0.32 |  | 75–50 | 0.53 |  | 20–10 | 0.00 |
|  | 400–300 | 0.00 |  | 300–250 | 0.03 |  | 200–150 | 0.23 |  | 50–25 | 0.16 |  | 10–0 | 0.00 |
|  | 300–250 | 0.00 |  | 250–200 | 0.00 |  | 150–100 | 0.10 |  | 25–0 | 0.47 |  |  |  |
|  | 250–200 | 0.07 |  | 200–150 | 0.04 |  | 100–80 | 0.00 |  |  |  |  |  |  |
|  | 200–150 | 0.37 |  | 150–100 | 0.10 |  | 80–60 | 0.00 |  |  |  |  |  |  |
|  | 150–100 | 1.23 |  | 100–50 | 0.25 |  | 60–40 | 0.09 |  |  |  |  |  |  |
|  | 100–50 | 0.33 |  | 50–25 | 0.49 |  | 40–20 | 1.53 |  |  |  |  |  |  |
|  | 50–0 | 0.10 |  | 25–0 | 1.25 |  | 20–0 | 5.71 |  |  |  |  |  |  |
| 7 | 600–500 | 0.00 | 12 | 130–100 | 0.00 | 16 | 80–60 | 0.48 | 20 | 80–75 | 0.90 | 26 | 30–20 | 0.12 |
|  | 500–400 | 0.00 |  | 100–50 | 0.00 |  | 60–40 | 0.00 |  | 75–60 | 0.09 |  | 20–10 | 0.03 |
|  | 400–300 | 0.01 |  | 50–25 | 0.00 |  | 40–20 | 0.46 |  | 60–40 | 0.16 |  | 10–0 | 0.03 |
|  | 300–250 | 0.00 |  | 25–0 | 0.00 |  | 20–0 | 0.00 |  | 40–20 | 0.07 |  |  |  |
|  | 250–200 | 0.00 |  |  |  |  |  |  |  | 20–0 | 0.00 |  |  |  |
|  | 200–150 | 1.08 |  |  |  |  |  |  |  |  |  |  |  |  |
|  | 150–100 | 1.57 |  |  |  |  |  |  |  |  |  |  |  |  |
|  | 100–50 | 0.00 |  |  |  |  |  |  |  |  |  |  |  |  |
|  | 50–0 | 0.61 |  |  |  |  |  |  |  |  |  |  |  |  |
| 9 | 900–700 | 0.00 | 13 | 450–400 | 0.05 | 17 | 140–120 | 15.45 | 21 | 40–20 | 0.05 |  |  |  |
|  | 700–600 | 0.00 |  | 400–300 | 0.11 |  | 120–100 | 15.44 |  | 20–0 | 0.07 |  |  |  |
|  | 600–400 | 0.00 |  | 300–250 | 0.10 |  | 100–80 | 0.08 |  |  |  |  |  |  |
|  | 400–300 | 0.00 |  | 250–200 | 0.26 |  | 80–60 | 0.00 |  |  |  |  |  |  |
|  | 300–200 | 0.00 |  | 200–150 | 2.80 |  | 60–40 | 0.00 |  |  |  |  |  |  |
|  | 200–140 | 2.41 |  | 150–100 | 1.11 |  | 40–30 | 0.00 |  |  |  |  |  |  |
|  | 140–100 | 2.77 |  | 100–50 | 0.04 |  | 30–20 | 0.06 |  |  |  |  |  |  |
|  | 100–50 | 2.17 |  | 50–25 | 0.00 |  | 20–10 | 0.12 |  |  |  |  |  |  |
|  | 50–0 | 0.29 |  | 25–0 | 0.00 |  | 10–0 | 0.27 |  |  |  |  |  |  |

*Eukrohnia hamata*

| Station number | Depth (m) | Ind·m$^{-3}$ | Station number | Depth (m) | Ind·m$^{-3}$ | Station number | Depth (m) | Ind·m$^{-3}$ | Station number | Depth (m) | Ind·m$^{-3}$ | Station number | Depth (m) | Ind·m$^{-3}$ |
|---|---|---|---|---|---|---|---|---|---|---|---|---|---|---|
| 5 | 200–160 | 0.04 | 10 | 600–500 | 0.22 | 14 | 500–400 | 0.12 | 18 | 180–160 | 0.00 | 22 | 30–20 | 0.00 |
|  | 160–140 | 0.00 |  | 500–400 | 0.29 |  | 400–300 | 0.02 |  | 160–120 | 0.00 |  | 20–10 | 0.00 |
|  | 140–115 | 0.00 |  | 400–300 | 0.20 |  | 300–250 | 0.00 |  | 120–100 | 0.00 |  | 10–0 | 0.00 |
|  | 115–100 | 0.00 |  | 300–250 | 0.32 |  | 250–200 | 0.02 |  | 100–80 | 0.00 |  |  |  |
|  | 100–80 | 0.00 |  | 250–200 | 0.19 |  | 200–150 | 0.00 |  | 80–60 | 0.00 |  |  |  |
|  | 80–60 | 0.00 |  | 200–150 | 0.14 |  | 150–100 | 0.00 |  | 60–40 | 0.00 |  |  |  |
|  | 60–40 | 0.00 |  | 150–100 | 0.13 |  | 100–90 | 0.00 |  | 40–20 | 0.00 |  |  |  |
|  | 40–20 | 0.00 |  | 100–50 | 0.00 |  | 90–50 | 0.00 |  | 20–10 | 0.00 |  |  |  |
|  | 20–0 | 0.00 |  | 50–0 | 0.00 |  | 50–0 | 0.00 |  | 10–0 | 0.00 |  |  |  |
| 6 | 500–455 | 0.55 | 11 | 500–400 | 0.93 | 15 | 434–300 | 0.08 | 19 | 106–75 | 0.00 | 23 | 30–20 | 0.00 |
|  | 455–400 | 0.79 |  | 400–300 | 0.74 |  | 300–200 | 0.00 |  | 75–50 | 0.00 |  | 20–10 | 0.00 |
|  | 400–300 | 0.32 |  | 300–250 | 0.13 |  | 200–150 | 0.00 |  | 50–25 | 0.00 |  | 10–0 | 0.00 |
|  | 300–250 | 0.43 |  | 250–200 | 0.18 |  | 150–100 | 0.06 |  | 25–0 | 0.00 |  |  |  |
|  | 250–200 | 0.04 |  | 200–150 | 0.17 |  | 100–80 | 0.00 |  |  |  |  |  |  |
|  | 200–150 | 0.12 |  | 150–100 | 0.09 |  | 80–60 | 0.09 |  |  |  |  |  |  |
|  | 150–100 | 0.05 |  | 100–50 | 0.04 |  | 60–40 | 0.00 |  |  |  |  |  |  |
|  | 100–50 | 0.00 |  | 50–25 | 0.00 |  | 40–20 | 0.00 |  |  |  |  |  |  |
|  | 50–0 | 0.00 |  | 25–0 | 0.00 |  | 20–0 | 0.04 |  |  |  |  |  |  |
| 7 | 600–500 | 0.58 | 12 | 130–100 | 0.03 | 16 | 80–60 | 0.00 | 20 | 80–75 | 0.00 | 26 | 30–20 | 0.00 |
|  | 500–400 | 0.07 |  | 100–50 | 0.02 |  | 60–40 | 0.00 |  | 75–60 | 0.00 |  | 20–10 | 0.03 |
|  | 400–300 | 0.00 |  | 50–25 | 0.00 |  | 40–20 | 0.00 |  | 60–40 | 0.00 |  | 10–0 | 0.01 |
|  | 300–250 | 0.38 |  | 25–0 | 0.00 |  | 20–0 | 0.00 |  | 40–20 | 0.00 |  |  |  |
|  | 250–200 | 0.33 |  |  |  |  |  |  |  | 20–0 | 0.00 |  |  |  |
|  | 200–150 | 0.20 |  |  |  |  |  |  |  |  |  |  |  |  |
|  | 150–100 | 0.04 |  |  |  |  |  |  |  |  |  |  |  |  |
|  | 100–50 | 0.04 |  |  |  |  |  |  |  |  |  |  |  |  |
|  | 50–0 | 0.03 |  |  |  |  |  |  |  |  |  |  |  |  |
| 9 | 900–700 | 0.27 | 13 | 450–400 | 0.09 | 17 | 140–120 | 0.04 | 21 | 40–20 | 0.00 |  |  |  |
|  | 700–600 | 0.20 |  | 400–300 | 0.04 |  | 120–100 | 0.03 |  | 20–0 | 0.00 |  |  |  |
|  | 600–400 | 0.04 |  | 300–250 | 0.00 |  | 100–80 | 0.00 |  |  |  |  |  |  |
|  | 400–300 | 0.00 |  | 250–200 | 0.00 |  | 80–60 | 0.00 |  |  |  |  |  |  |
|  | 300–200 | 0.25 |  | 200–150 | 0.00 |  | 60–40 | 0.00 |  |  |  |  |  |  |
|  | 200–140 | 0.48 |  | 150–100 | 0.02 |  | 40–30 | 0.00 |  |  |  |  |  |  |
|  | 140–100 | 0.00 |  | 100–50 | 0.00 |  | 30–20 | 0.00 |  |  |  |  |  |  |
|  | 100–50 | 0.00 |  | 50–25 | 0.30 |  | 20–10 | 0.00 |  |  |  |  |  |  |
|  | 50–0 | 0.00 |  | 25–0 | 0.00 |  | 10–0 | 0.00 |  |  |  |  |  |  |

Druck:          Strauss Offsetdruck, Mörlenbach
Verarbeitung:   Schäffer, Grünstadt